THE
ILLUSION OF
TIME

Michael Berossa

authorHOUSE®

AuthorHouse™ UK Ltd.
500 Avebury Boulevard
Central Milton Keynes, MK9 2BE
www.authorhouse.co.uk
Phone: 08001974150

First published by AuthorHouse 4/28/2008

ISBN: 978-1-4343-5975-9 (sc)

Printed in the United States of America
Bloomington, Indiana

This book is printed on acid-free paper.

Contents

TABLES & CHARTS

FIGURES

Introduction

This short book will re-explain scientific observations of the large and small-scale universes in terms of a theory that intrinsically links time and gravity as one phenomenon. It goes on to explain all atomic phenomena in the new paradigm. The broad precepts of cosmology, including Special and General Relativity theories, survive largely unchallenged, but a fundamentally new way to explain gravity will turn several standard beliefs on their head.

Although the Standard Model is regularly bombarded by "alternative" theories, this is simply a feature of scientific enquiry. Even Hubble – to whom the discovery of cosmological red shift is attributed – did not believe it actually indicated the recessional motion of galaxies. He thought it to be an effect only giving the same appearance, and personally preferred a universe with a mild expansion force, to balance inter-galactic gravitational forces. He believed that red shifts were produced as some kind of linear vacuum effect.

A grasp of physics will certainly be helpful for those reading and evaluating this book. Media exposure nowadays helps the public to stay informed of the latest research focuses in atomic physics and cosmology. The mathematical subtleties of theories are generally avoided, as the layperson is more interested in the ideas than the details; the layperson assumes the scientist is doing his sums correctly. Documentaries have done a great deal to educate the public about Special Relativity and its strange effects in extreme physical situations.

Relativistic effects like time dilation, length contraction and mass increase have all been verified experimentally, and can be understood mathematically from the relativity theories. The effects arise because of a need for invariance of measured quantities in different reference frames. While these effects are predicted mathematically, the uncomfortable truth is that most physicists would probably not naturally predict these effects on a physical basis alone. The theory you will read here has the advantage that it does just that. It will call for reinterpretation of the evolution of the universe, the nature of Black Holes, the cause of the Big Bang, the value of Ω, theories of galaxy formation and the true reason for cosmological red shift.

This largely discursive work starts with a central idea about the true nature of time, and shows the consistency of this idea on the largest and smallest scales possible. Numbers and equations inevitably crop up in several key sections, as the physicists among us quite reasonably need numbers to be convinced. Some mathematical substantiation therefore remains inescapable, but a good High School background in mathematics will be enough for anyone to cope.

The theory starts from the demonstrated link between clock rates and gravity. The central thesis is that time is a measure of the *net flux* of gravitons from all atoms. Gravity is taken to be a form of radiation, and the radiation pressure associated with gravity is a force of expansion, driving the expansion of the universe. Gravitational attraction is a side effect of a radiation effect that also allows absorption, and spatial expansion and mass-to-mass attraction become two facets of just the same phenomenon. Discussions concerning time and gravity also have consequences for understanding photon frequency, and a better explanation for cosmological red shift will emerge, with a new way to measure distance, age and expansion in the universe. Once red shift is understood properly, the reader will see that the universe is expanding at no more than half the rate believed.

In a discussion about the radiation properties of the atom, an important new property of matter emerges – in this book called the "critical radiation density". This property suggests that matter spontaneously decomposes into energy under critical radiation conditions. Radiating matter reaches this critical level well before it reaches that matter density that results in a Black Hole, and this in part argues for the non-existence of Black Holes. The same property seems to account for the high radiation of quasars: they are exploding gas clouds, in which graviton flux from atoms has become negative in value, and whose debris form the first galaxies. Critical radiation density – applied internally to the atom – also appears to account for the observed mass-to-radius relation of the neutron. The critical mass density usually discussed in connection with a Black Hole seems better applied to the miniscule pile of gravitons confined deep at the heart of the atom.

The quantum character of the vacuum, and the quantization of sub-atomic processes, is assumed. A quantum surface-velocity of at least 10^{13} m/s is also argued for, and dimensional analysis suggests that 10^{27} m/s and 10^{18} m/s are other possible values. High surface speeds in the vacuum would resolve certain quantum paradoxes, and bring about an elegant synthesis between Quantum and Classical Physics.

Allowing a General Relativity (GR) Theory to apply to the whole universe is an approach still open to debate. The universe (if not galaxies) is understood to be expanding everywhere at the speed of light. It has always done so, except for a short period near the onset of the quasar era. The GR universe described in the standard teaching institutions has no inside or outside, and consequently there is no "expanse" into which the expanding universe is considered to expand. Such a universe may seem rather mystical, but this description is just a faithful generalisation of the Special Relativity (SR) Theory, which disallows a local universe to have a discontinuous boundary. This is a mathematical result, even if it may fall short of the reality.

In the universe of this new model, the movement of gravitons and photons in the vacuum *defines* the expanding space, so that they move in a kind of "free fall" in a general expansion at speed c, although at the same time pushing against each other. The need for a mechanism to propagate photons and gravitons is largely circumvented, yet Maxwell's electromagnetic theory for light still works as a mathematically equivalent model. This new model – in the same way as the old - sees space "curve" near massive or high-speed objects, and allows energy-density changes in the vacuum.

The Standard (Cosmological) Model argues that the gravitating universe *should* be decelerating after the Big Bang, as it is apparently controlled only by its self-gravity. A decelerating universe has a chance that it might eventually "close" (and so see the initial expansion followed by a later crunch), but a decelerating universe could also be "flat" or even "open". Theorists prefer a critically (i.e. only just) closed universe, so struggle with latest observational evidence that the universe may still be accelerating (and therefore unequivocally "open"). Some papers lately have tried to bring back Einstein's original acceleration parameter, which was rejected early on in the theory of the Big Bang. The model to be presented here will describe a universe that expands at a fixed rate (thus appears "open" now) but that in its last days will finally decelerate to a halt, as matter finally expires.

A few redefinitions of fundamental quantum sizes and times have been used in this discourse. In the standard literature, these are dimensionally defined quantities – that is, they are "natural" quantities based on combinations of G, c, h, and unit-less constants. Thus, the "Planck mass" can be defined by simple combinations like $\sqrt{hc/G}$ or $\sqrt{hc/2\pi G}$, or any combination of physical constants (with their associated units) that has an overall unit of only mass. The "Planck length" is the distance a particle can cover if it travels at the speed c for one unit of the Planck time. The "Planck time" seems to be a number like $\sqrt{hG/2\pi c^5}$, which is equal to 5.39×10^{-44} seconds, and the Planck length is consequently formed from $\sqrt{hG/2\pi c^3}$, which represents a length of around 1.6×10^{-35} meters. This differs from the Planck radius, which is around c^2 times greater, at 1.64×10^{-18} meters.

The other two important sizes in any discussion are the Compton and gravitational radii. The Compton radius is approximately the size of the atom, as measured in the external universe. It represents a length at which scattering from an atomic surface takes place. Its size is around 2.8×10^{-15} meters. The gravitational radius – also called the Schwarzschild radius – is the critical radius at which matter becomes self-confined beyond retrieval. It appears in discussions about Black Holes, but also in discussions about the density of material at the very core of the atom. It is defined by the ratio $2Gm/c^2$, which – for an atom – amounts to the size $2 \times 6.67 \times 10^{-11} \times 1.96 \times 10^{-27} / 9 \times 10^{16} = 2.91 \times 10^{-54}$ meters, which is about 40 orders of magnitude smaller than the Compton sphere.

The fundamental quantum length used in the discussions to follow will, however, be defined more simply as h/mc (about 10^{-15} meters), and the "Planck length" to be used in calculations in this book will be defined as the geometric mean of this fundamental quantum length and the standard gravitational radius – that is, as $([mG/c^2][h/mc])^{\frac{1}{2}} \approx (10^{-53} . 10^{-15})^{\frac{1}{2}} = 10^{-34}$ meters. The fundamental length h/mc similarly defines a Planck time of $10^{-15}/c \approx 10^{-23}$ seconds.

As a fun "overlay" to the central themes, the chapters begin with historical references that sound like interventions of flying saucers in human affairs. The UFO phenomenon is an enigma for which I do not pretend to have an answer, but the captions provided suggest it is not a recent phenomenon. UFOs seem to span human history. Maybe the medieval interpreters were right when they supposed UFOs were vehicles used by angels or spiritual watchers over the human race. Or perhaps von Daniken is right in thinking the "gods" were really astronauts, and the human race is just an ancient (or even ongoing) genetic experiment originally carried out by extraterrestrials. Any "super-technology" employed by extraterrestrials must nevertheless obey the same laws of physics that apply everywhere. With a new theory of time and space, it may be possible to say as well whether a UFO-like craft really is possible, and later chapters address this question for fun.

Thanks are due to certain persons who took the trouble to wade through the earliest versions of this book and point out where they lost their way. One person took my manuscript to read, and was never seen or heard from again. One bright young student of mine - Umar Agha - studied key sections of the book for logical or mathematical errors, and was most helpful in regard to showing me where the argument got hard to follow. Others got part way through, but didn't finish. At best, I think, the book is still difficult for the average reader, but - for the one who can access it – it will be a most fascinating read.

The cover has a spacecraft that looks like a timepiece – suggesting that the only practical UFO would have to be some kind of "time machine". The only time machine I can speak of with some confidence right now is the atom itself. We need to understand the atom properly first, before trying to build a craft that defies the clock. Hopefully what I have written will advance that understanding. Questions about ideas presented in this book are of course welcome from anyone, and can be directed to the author through the publisher.

WHY DOES THE UNIVERSE EXPAND AT SPEED C?

1450 BC: During the reign of the Pharaoh Thutmose III, there is a description of multiple "circles of fire", brighter than the sun and about 5 metres in size, that appeared over multiple days. They finally disappeared, after ascending higher in the sky.

Figure 1: Numerous UFOs over Nuremberg, Germany
(Hans Glaser wood-cut from 1566, five years after the event on April 14th, 1561.)

The expansion speed of the universe currently appears[1] to be the speed of light, c, although many theorists propose a much faster initial speed in the very early universe, as part of an "inflation" theory. This initial hyper-expansion is believed to have decelerated back down to c after a short time. The expansion speed derived from observing receding galaxies gives a slightly lower figure than c, and the most distant luminous objects – quasars – are found at cosmological recession speeds beyond $0.8c$. But these are not the first objects in the universe. Galaxies may on average recede with the cosmic expansion, but the current placement of the first galaxies at $z \approx 0.8$ can fit a model in which the universe is, say, 13.7 billion years old or one in which the production of galaxies spans from 13.7 – 16 billion years ago, and the universe is older still.

Plainly, free photons propagate in all directions at c, and this - on its own - implies a universal expansion of the vacuum at a speed no less than c. Latter-day theories involving "multiple universes", collisions of "cosmic (mem)branes", and more, all continue to incorporate the early super-luminary expansion, and this now appears entrenched in the thinking. Those who first proposed this early hyperinflation did so to explain the extreme uniformity of the cosmic microwave background radiation (CMB). However, the whole idea of inflation flies against the tenets and spirit of GR - besides being beyond observational verification. CMB smoothness can be explained more easily in the current theory, and does not lean upon some pre-relativity inflation theory.

We are compelled to believe that the empirical evidence for the expanding universe comes to us directly in the form of Doppler red-shifted light from receding galaxies. The Hubble Law – which directly links red shift to distance – came about by combining a red shift-to-surface-brightness correlation with a surface brightness-to-distance correlation,

---

[1]　Cosmologists, on the whole, would probably not be happy to allow such a glib statement, preferring to leave that issue open, and basing their expansion speed purely on maximum observed red shift as lower bound. It also goes against the view that the expansion rate could be varying with time (i.e. either accelerating or decelerating). Nevertheless, the Standard Model is prone to a certain circularity of thought, as it presumes that red shifts indicate recession speeds of galaxies in what they infer to be an expanding universe, but that the galaxies have these recession speeds mainly because of the expansion of the universe.

to arrive at a red shift-to-distance law. Difficulties with galaxy luminosity functions, and with *peculiar* motions superimposed over cosmological recession velocities, mean that this Law's proportionality constant - the Hubble constant - remains about 10% uncertain.

Although seen "now", galaxies are always seen at earlier moments in the history of the universe. A look-back "horizon" is anticipated, where – if the universe is not too old - the observer should see back entirely to events at the origin of the universe. But it should be remembered that look-back time and distance are always estimated on a *local* clock.

The universe has a *natural* observation horizon for another reason. Once cosmological recession speeds go beyond c, light can no longer find a path back to the local observer. The receding object then disappears "over the horizon". Another key point is that all views "out" from the local frame are, in reality, views looking "in" on an earlier, smaller universe. The observer's viewing "edge" will necessarily be where the first visible state of the universe is seen, near its start-radius. No look-back, of course, could see beyond this. Views in *all* directions must converge on this event, if the universe is a GR entity.

The observational limit is also defined by where Doppler recession velocities reach c. If a receding metric does not recede at more than c, the distance ct at which this limit is encountered must be the same as that to the creation of the universe. Astronomers struggle to find objects beyond $z = 0.8$ – about 80% of the way to the creation event. This inability could be due to two things: the universe may not have *started* with a Big Bang (this may have been a later event within the evolution of the universe – say, at $z = 0.8$), or else material beyond $z = 0.8$ is just too dark to see.

The expansion of the universe at constant speed c means that all local lengths, however small or large, expand by the same factor dR/cT in the relative time dT/T, where dT is the duration of the observation and T is the current age of the universe. Because T is on the order of 10^{10} years, here and now, all lengths must increase at the relative rate of about 1 part in 10^{10} per year, or 2.65 parts in 10^{18} per second (for a universe of age T = 12 billion years).

If the universe most resembles an inflating balloon - i.e. an object re-scaling itself around some fixed centre - then the horizon expansion at c would be the cumulative effect of all local expansions. At all shorter radii, the expansion rate must be some fraction of the horizon velocity. The GR model does not implicitly negate the expanding balloon idea. One could argue that a local observer is simply at that radius in an expanding balloon model where – out to the local horizon - every process happens to be limited to a maximum velocity c. The *actual* boundary could – in this model - be extending at some greater speed. You would not know the difference.

The only difference would be that observers looking in different directions would not be able to see a single creation event at their local horizon. Lorentz transforms should certainly still work locally, and red shifts would still seem to show an even *recession* in all directions from the observer, given that the horizon would then define a "local" universe. Also, $\Delta\lambda/\lambda$ would just map $\Delta c/c$ in any radial direction. Although this is slightly different to the true Doppler relation ($\Delta\lambda/\lambda = \Delta v/c$), if the local value of c is just some "v" with respect to a greater value of c at a greater radius, it amounts to the same thing.

Therefore, a single-centre expansion *could* conceivably be a workable model, except that it must allow c to vary with radial distance. The dT/T perception is the same in this and the standard GR model. The principle objection to the "balloon" model is that the assumptions underlying GR - namely, that space-time is a single, geometric, self-defining object, not expanding with respect to anything except its own vacuum state - would be violated, and GR could break down at large scales.

The same criticism would arise in the scenario where a "cosmic fountain" filled a "central" void with new material emerging everywhere and at all times with speed c. Within a truly GR universe, this *fountain* could be considered to be everywhere - not necessarily at a specific centre.

One point is as much a centre as another. The vacuum would simply have a fixed rate at which it produced new matter, linked to the rate at which the volume of the universe was increasing, and this would be a universal effect. This type of model suffers from the need to explain how matter continues to be created in this way, and how the Conservation of Energy principle would work.

An explosion scenario (the way the "Big Bang" is commonly understood) has many of the same geometric features as a GR universe, although the universe would have some sort of "hollow" (if unobservable) centre, rather like the hollow shells seem around supernova remnants. The universe itself would then be just an expanding shell of material. Such a shell – as its radius grows – must grow *relatively thinner* with time in one dimension, but appear infinite (indeed, wrapping back around on itself) in the perpendicular plane to the flow. Strange boundary effects would be evident, because of the asymmetric mass distribution around an observer. The universe does not appear to be like this either.

General Relativity modifies the balloon picture by insisting that *every* point acts as if it were the centre around which the inflation occurs. This stems from an implicit constraint that the universe should "look" the same at all times and in all places to all observers - a requirement for applying a *general* relativity theory. General Relativity is in the first instance a framework designed to handle measurement and time, and is concerned with doing repeatable physics anywhere, now and in all epochs.

This sameness in all places and times is called the Cosmological Principle. That everyone shares the illusion of being located at the "centre" of the universe is taken the one step further of arguing that everyone *is* - as legitimately as anyone else – at the centre. This is equivalent to saying the universe has no centre at all. And - if no centre - then also no boundary! GR will therefore not tolerate a boundary condition. The universe has no "edge'. The observed expansion of the universe is not therefore due to an "explosion" in any normal sense of the word. Explosions have centres, but the universe expands from everywhere.

This elegant model seems puzzling to the layman, because it asserts that the universe expands at *every* point with speed *c,* yet the observed *net* expansion is also at speed *c.* Yet this is where the Lorentz transforms, in the context of Special Relativity, play their part, showing themselves true in any local universe, and also in the sum of all local universes. The consequence of such an approach is the *simpler* mathematical notion that the universe is, and has been, expanding with the *same* speed *c* at *all* points. The generalized Lorentz transforms also shorten the "distant" universe (which amounts to slowing its clocks) at any place.

A nice feature of the Lorentz transforms is that they are purely geometrical transforms, whether or not $c = 3 \times 10^5$ km/s. They are *geometric*, rather than empirical, relations. Incorporated into General Relativity, the transforms will work for *any* value of *c* that might be plugged into the equations. If *c* were a time-varying quantity, GR would simply exchange *c* for $c(t)$, and a few extra derivatives would appear in certain calculations.

The Lorentz transforms were around a long time before the idea of an expanding universe, but they tolerate a relatively slow expansion of the universe in their generalisation. They deal accurately with systems in which gravity (specifically, acceleration) remains a small effect compared to all other inertial and electromagnetic (EM) forces. The transforms legitimately apply to material objects while $v < c$, but for non-material bosons only while $v = c$. This makes it illegitimate to use the reference frame of the boson to describe material reality, and vice versa. A quantum of radiation cannot be isolated in the material world, and it does not perceive that world. Each type of object has its own reality.

To have a space that is "geometrical" at *all* scales in GR actually requires the expansion of the universe to be at speed *c,* and the equations prefer a "critically closed" solution for the universe as a whole. A metric that recedes at more than *c* makes the universe become causally detached at large scales, so that it can no longer be described as being one metric. The equations then need modifying, as seen in the Minkowski metric. So it would be no accident at all if the speed of expansion of the universe turned out to be the speed of light exactly, as this is the speed that appears in the transform, and this would allow the universe to be treated at all times as a single measuring system. A change in the speed of light would translate into a change in the universal speed of expansion, and it seems true to say that it is both the speed of the expansion that sets the speed of light, and vice versa.

GR will not work at the quantum level, however, as the vacuum is presumed to be a quantum medium, whereas GR argues from the perspective of an infinitely divisible medium. With these presumptions, the best that can be done theoretically is to define a scale at which the transition between approaches should be made. Yet perhaps the question of whether quantum states arise as an intrinsic property of the vacuum rather than as some kind of partitioning of the vacuum brought about by the presence of material particles – or whether, maybe, it arises for *both* reasons – needs to be answered first. Certainly the atomic radius is loosely that dimension at which the changeover occurs[2], and a change in properties should occur on either side of this radius. Near this radius, GR and quantum mechanical equations should nearly agree.

Special Relativity predicts that space shrinks and clocks run slower, when the local relative velocity v approaches the speed of light. Somewhere near $v = c$, space and time approach quantum dimensions. In the rest frame of the observer, quantum statistics and selection rules necessarily take over as the correct way to describe the physical situation. Since transformation between reference frames is impossible in quantum physics, the formalism of GR breaks down at the ultra-relativistic stage. This logical dilemma (which, however, can also be considered to be the *real* reason that physical objects cannot travel at the speed c) indicates that the boundaries of space and time can be described only in the language of quantum physics, and not geometry. Both SR and GR are curtailed as $v \to c$.

---

[2]   Although theorists often push this "changeover" radius right down to the Planck length.

HOW DO LIGHT AND FORCE PROPAGATE THROUGH SPACE?

Figure 2: Flaming Girder
1465AD - A flaming girder is seen in the sky during the reign of Enrico
IV. (From *Notabilia Temporum*, by Angelo de Tummulillis.)

The preceding discussion may have removed the sting from the question of why the speed of light and of the universal expansion should in principle and preferably be the same, but it leads in another interesting direction. It suggests that any radiation – if regarded quantum-mechanically as being constituted of vacuum particles - can itself only be in a kind of free-fall in the cosmic expansion. That is, light does not so much move *through* the universe as *with* it, and in reality there may be no relative velocity at all. It may not be necessary to credit space with properties and mechanisms for propagating radiation.

Before GR or Quantum Mechanics (QM) appeared, observations involving diffraction and polarisation fostered a strong belief in light as an electromagnetic wave. This wave evidently *propagated* in a smooth, stationary vacuum "medium". In the model presented here, the vacuum is not "fluid" in this way, and vacuum particles do not share elastic bonds in the way that atoms do in a fluid or solid. Transverse wave propagation, from this point of view, should not be possible, and the fact that no experiment (except perhaps the Einstein-Rosen-Podolsky trial) can show any motion that can *overtake* the general expansion (like someone walking on an escalator) suggests that space is, by nature, *inert* to motion. The motion of photons in any local reference frame simply shows that space itself is expanding through any reference frame at the speed of light, and photons sit in and partly contribute to this flow.

The observation that a photon behaves *as if* it had a wavelength and frequency derives simply from its *intrinsic* oscillation. Polarisation experiments confirm that light always has a plane of oscillation. The intrinsic oscillation frequency is perceived as the photon's frequency, and photon wavelength is just the rotation of the photon mapped in vector addition to the velocity of the photon in the vacuum. The photon, however, simply sees itself as spinning or oscillating at a single spot, unaware of any translation. Indeed, the photon is a stationary state. But in *projection*, in the observer's frame, a wave-like object is observed.

The photon certainly also has an "electric" character – or, at least, responds to the presence of atoms in its vicinity, as light interference patterns result from an interaction between photons and the electrons of the atoms forming the edge of a diffraction slit. Even in refraction, frequency is conserved, even if c appears reduced. A photon itself cannot be altered by the outside world, unless conditions exist allowing the photon to be energized or de-energized. De-excitation means removal of angular momentum from a photon, but this does not remove the intrinsic spins of those elements that form the photon. The photon "disappears" when its angular momentum is totally transferred, and an atomic system correspondingly receives and stores this energy as orbital and/or rotational energy of an electron, causing an excitation state for that atom. Energy reduces to nothing more than a measure of the total angular momentum of a system, as recorded in the algebraic sum of the rotation states[3] of all particles involved, including the rotation states of local gravitons.

A photon's energy is classically defined as both $E = hc/\lambda$ and $E = hf$, and entrance into a denser refractive medium sees a photon's wavelength shorten in proportion to the speed reduction. The oscillation frequency does not alter – as this is specific to the photon, and invariant. The equation $E = hf$ therefore records no change in energy, even in the new medium, unless the value of Planck's constant, h, changes. But a change in h should be semantically impossible, as h is the defining quantum number of action, which should be universal and invariant. If h remains constant, the alternative equation $E = hc/\lambda$ needs the *ratio* c/λ to stay constant to conserve energy. And this is what is reported in refraction experiments.

Only in this one respect, then, could the speed of light change, depending upon the refractive index, n, of the transmitting material. Refraction will be discussed again near the end of this book, where even the notion that c changes in dense media will be questioned. In free space, in any case, n remains essentially 1, so this is not a factor that generally needs handling in what follows.

Gravitons are bosons[4] that mediate gravitational force, while photons mediate EM[5] forces. These are the only two physical forces with zero rest-mass exchange particles. All other physical forces are transmitted by bosons with finite rest masses and finite lives. Force-transmitting bosons operate over the extremely short, sub-atomic distances between the respective fermions that emit and absorb them, and exist for only a few quantum moments. Cosmological effects, however, are entirely explicable in terms of zero-rest-mass particles that are not extinguished at all.

---

[3] Spin and rotation are different. Spin is intrinsic, but rotation is extrinsic. A photon has both intrinsic spin and rotational excitation states, directed in orthogonal planes.

[4] So named because they follow Bose-Einstein statistics, linked to spin-pairing.

[5] EM means "electro-magnetic".

TIME

... in Tarquinia towards sunset, a round object, like a globe, a round or circular shield, took its path in the sky from west to east."

- *Book of Prodigies,*
Julius Obsequens – re: events in 99BC

At Aenariae, while Livius Troso was promulgating the laws at the beginning of the Italian War, at sunrise, there came a terrific noise in the sky, and a globe of fire appeared burning in the north. In the territory of Spoletum a globe of fire, of golden color, fell to the earth gyrating. It then seemed to increase in size, rose from the earth and ascended into the sky, where it obscured the sun with its brilliance. It revolved toward the eastern quadrant of the sky.

Julius Obsequens, re: events in 90 BC

Something like a sort of weapon, or missile, rose with a great noise from the earth and soared into the sky.

Julius Obsequens, re: events in 42 BC

Time is psychologically understood because of ageing. All things made of matter age. Ageing is irreversible, so physical processes always run in the same direction. Laws of Thermodynamics encapsulate this observation in the physical world in terms of the maximisation of entropy. That is, systems like to dissipate energy. A few high-energy photons will eventually be replaced by a larger number of low-energy photons. The universe expands, and cools. Replicating systems fail to replicate perfectly, and tend not to correct faulty replications. Thus, species mutate and degenerate. Atoms – once formed – slowly decompose through radioactive decay and graviton emission. These processes, without intervention, are always one-way.

One visualizes time as an unbroken line or as a propagating plane in space, similar to a physical path traced through space. Mathematicians are used to drawing diagrams with a directed time axis, as if time had a "direction". Time, of course, has no direction. It is an abstract quantity, and measures something. Time, in fact, is a number. We assign the descriptor of "rate" to a clock, and this is to clarify that the number described as time is equivalent to some "rate" at the atomic level. This same "rate" also determines the strength of a gravitational force.

Nor is time smooth. Just as the volume of space is divided microscopically into discrete quantum locations, so – at the highest energies, according to quantum equations - time begins to divide into quantum units. It becomes grainy and countable. The secret to understanding time is to see it from the point of view of *one* atom, and to regard the atom as an aging entity. If this ageing contributes to the dynamical history of the universe, it means that atoms themselves become the *clocks* of the universe, and time in reality measures a discrete atomic quantity. This transfigures time from an abstract to a *physical* quantity.

It seems almost obvious that time should track the emission rate of an atomic particle common to all matter. And the simplest such particle is the graviton. Another strong reason to single out the graviton from all other choices is the experimental evidence that local gravitational forces are directly responsible for altering the rate at which time flows. Quantum theories of matter and energy present matter as existing in various states of tension, produced by particles that transport *forces* from one place to another. The atom, by experiencing gravitational force, is naturally a stressed entity. Time therefore indirectly measures the stresses experienced by an atom.

In the current theory, gravitons are more than "exchange" particles – they actually *originate* in atoms, where a vast store of their constituent parts is found. These constituent parts exist under high compression – indeed, probably at or near critical mass density - deep within the nucleus of each ordinary atom. Under normal conditions, an object arises from deep within the Planck radius, which is eventually expelled from the atom as a graviton. This is a purely quantum process.

Quark action is implicitly involved in the mediation of this process, as the systematic expulsion of gravitons seems to be a natural process in all atoms. This regular emergence of gravitons from an atom is what is meant by "clock", and net emission - if it could be measured - would represent a way of measuring or quantifying time. Conversely, the local time rate indicates the local net rate of graviton emission.

The link between gravity and time follows naturally. A graviton is released and occupies a specific volume in the vacuum. It has a natural free size and a natural tendency to repel other gravitons. The emission of gravitons – that is, *gravity* – is directly responsible for the gentle expansion of the universe. The universe does not expand as an accident of broken symmetry: rather, it is the natural emission and expansion of gravitons by all material particles that drives this. Any radiation-based universe would expand – broken symmetry or not. Clearly, if gravitons emerge from the nuclei of atoms at speed *c,* the net effect of countless trillions of such emissions in all directions will be twofold. First of all, the general net expansion of space in *any* direction will be at speed *c.* Secondly, any local clock will run at a rate linked to the degree to which graviton emission remains unimpeded.

This may all seem to fly in the face of the concept that gravity is an *attracting* force. But if gravitons crowding into space produce expansion, the *negative* of this expansion – that is, the attractive force of gravity - must come about as many atoms also *absorb* gravitons from the surrounding space. Atoms remove discrete amounts of volume from space every time they assimilate gravitons arriving from other atoms. This appears to be a highly probable phenomenon whenever a graviton makes anywhere near a direct hit on an atom. Gravitons appear to assimilate *easily* and randomly into matter, even if their release is more regulated. A balance between emission and absorption must necessarily appear in any region containing many atoms, and the average *net* rate then determines the "local clock" rate.

The implications for time are interesting. When more matter is present, and especially if it is densely packed, the recapture rate of gravitons must climb, and does so in accordance with the relative fraction of the capture cross-section of space taken up by the cross-sections of material particles. Where more matter is present, there is both more re-absorption of gravitons (i.e. more gravitational force) and a lower net contribution to the local expansion of space. This acts like a refractive index for the local expansion, and the vacuum becomes denser near the gravitating mass. Thus, space curves in around massive or dense objects. An increased refractive index is ordinarily also measured in terms of a slow-down in the speed of light. Since the speed of light must remain constant to all observers in the vacuum, it follows that it is *time* that must slow in the same proportion, to preserve the local measure of c intact.

This makes sense, within the physical description of the situation. After all, if the *net* rate at which free gravitons appear is *time*, this is reduced in the presence of matter. The maximum emission rate is achieved when *no* other matter is present, and is tied inexorably to c. Thus, the way time rates vary is accurately described by Lorentz transforms. Time would *stop* (although this is actually disallowed for fermions) if atoms were located so closely together that the *net* rate of production of gravitons should fall to zero. Emission would equal re-absorption. This would be a critical radiation density for matter, and a critical state in time.

The larger the sphere over which net graviton emission is being considered, the greater the eventual recapture probability of any graviton by another atom. At least this would be so if the matter density remained constant throughout the volume element being considered. This being the case, the *net* emission from concentric reference spheres should appear to fall with the increasing radius of the reference sphere, and the net rate of time should similarly decrease in larger enclosing spheres. Larger viewing spheres correspond to greater distances over which observations are made, and thus to larger look-back times. In looking across the universe at other objects, one always looks *back* in time, and also looks back to when time ran slower. This description implies that clocks should appear to slow down at larger look-back distances.

Continuing this argument, it seems that the clock rate should seem to fall to zero at the observing horizon. In principle, one could estimate the *time rate* at any particular distance by supposing a universal gradient from one's maximal local rate at the "centre" of the local universe, out to zero at any horizon. This function, for reasons to be discussed later, is not linear, but logarithmic. It is, however, practically linear. In a way, this is done with cosmological red shift (which will also turn out to be a flat logarithmic relation), as the increase of red shift with distance emulates the way time would also seem to run slower, according to the radial distance from the observer. The horizon foreshortens in a way not dissimilar to that in which relativistic motion foreshortens distance[6].

The key point in this regard, however, is that red shift maps time-rate with radial distance, and that this will not turn out to be an exactly linear mapping. If a photon arrives in a modern detector with a certain red shift, this indicates a clock difference between when it was emitted and when it was received. The size of the shift (if there are no Doppler contributions) indicates the total clock difference between emission and absorption, and the argument given above suggests that the distance travelled can be known directly from this clock difference.

Events seen taking place in distant objects will necessarily *seem* a little sluggish on latter-day clocks. Events seen on nearer objects will appear

---

6   But see "The Twin Paradox" (page 62) for a clarification of the similarities and dissimilarities of these two situations.

less so, and local events will occur at a regular rate. This "low-energy" appearance of the more distant events is attributed, in the Standard Model, to gradual loss of energy to the vacuum, but events originating in a denser universe will always look more sluggish now, because the measuring frame has clocks that now run faster.

Clock slowdown can also be interpreted from a Newtonian point of view as being due to the fact that the *net effect* of universal gravitation is to measure as zero locally, at the "centre" of the local universe. The "clock" therefore runs fastest here. If someone wanted to consider time as reflecting a universal potential gradient, they would have to *integrate* the time in accordance with the associated red shift of the arriving photon, and deduce the actual source distance from this.

In the section *Mass and Energy*, the appearance of the c^2 term in the equation $E = mc^2$ is discussed. In one possible approach, c^2 is treated as a pan-cosmic gravitational potential, and becomes the ratio of the cosmic expansion velocity, c, to a local *time gradient*, $1/c$ (measured in some appropriate time units per metre). This implies that a gradient of some sort must exist. In short, a *locally flat* cosmic potential of c^2 appears in all places, and – in keeping with the principle of Conservation of Energy - is a quantity that would also be conserved in all "local" reference frames. The "gradient" is a universal effect.

Experiments with atomic clocks at different altitudes confirmed that clocks ran at different rates for different local gravitational potentials. It is usually said that gravity slows time, but really it is more accurate to say that any time rate is just a measure of local gravity, and vice versa. Time and gravity reciprocally measure the *same thing*, and this proves to be a useful result in the discussion to follow.

There is no such thing, though, as a *universal* or "right" clock, except in the sense that all clocks, in the absence of matter, will run at a rate correlated with the universe's speed of expansion c, which *is* a fundamental constant of nature. But from one locality to another in the universe, the conditions are different, so clocks run at different rates. As the Twin Paradox and other experiments prove, each measuring system in the universe *feels* time to be flowing at a fixed rate, regardless of its state of being or motion. It is the personal perception and subjection to time that produces ageing. Objects – when judged by a single local clock - age at different rates in the universe, following their various local clocks, but the sensation of ageing is always the same anywhere.

An important question arises as to how a boson experiences time. The Lorentz transform implies that an object travelling at speed c should see time stop. Or – better still – there is no concept of time at all. That is, any concept of change or ageing is lost when $v = c$ in Lorentz's transforms. The boson universe is frozen and static, and distances seem to be traversed in single quantum jumps. This is true regardless of how many times the electric spin vector may have cycled within a photon between its emission and re-absorption. The intrinsic oscillation is entirely unconnected to the concept of time. Vast geometric distances can seemingly be jumped in instances, and velocity and distance have no real meaning except as a description of position. The extremely localised universe of the photon does not appear to expand. It occupies a largely unique position in space until it finally interacts with another material particle, and fulfils its mission to deliver spin angular momentum (energy) and de-excite.

One should however exercise caution before linking distances in space to the alleged period that a photon has been *en route* between two locations. Certainly, distant objects will be seen in the universe in states younger in approximate proportion to their distance from the local reference frame. But the clock rate at the time of emission could have been *radically* different from that in the receiving frame at the same conformal time. The "distance" covered is necessarily defined in the receiving frame, and any inferred *age* is recorded in terms of the *current* clock on *Earth*. Any attempt to get a *history* of the universe just by looking at its distant objects must take into account the *masses* and densities of the objects being observed, as this could have affected the clock rates when the light left those objects. This would have to be factored in when establishing the distances covered by incumbent photons.

Space, Matter and Gravity

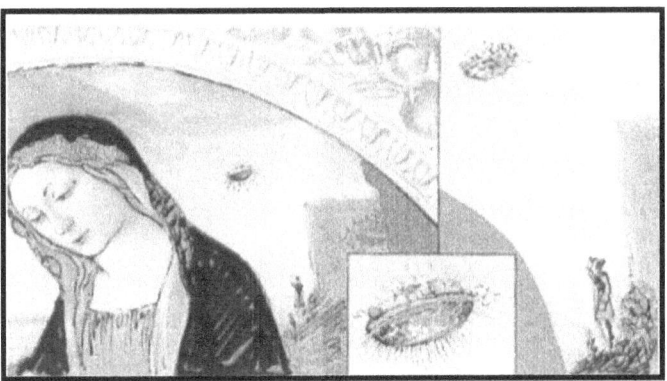

Figure 3: The Madonna
Flying saucer sightings were so common in the Middle Ages that scholars
began to think they were "angelic" craft, and artists began to incorporate
them into religious art. This sketch is of a painting called *The Madonna
with Saint Giovannino*, a painting by Domenico Ghirlandaio (1449-1494),
and hangs as part of the Loeser collection in the Palazzo Vecchio. Over
Mary's shoulder is a craft that - upon magnification - seems to show a
flying saucer with streams of radiation, and something like portholes.

Space – by which is meant the free vacuum - is needed both to produce
and absorb the universe. The whole universe, ultimately, has been
constructed from it. The solutions of GR, or indeed any other system,
have no other source for the energy found in matter than the vacuum of

space itself. Quantum Theory has, thankfully, clarified that the vacuum is *not* – as "common sense" might otherwise suggest – empty. The idea of something arising from "nothing" makes no ultimate sense. The vacuum expands and can be deformed through the presence of matter. A truly *empty* vacuum would not respond to mass with curvature. It would not expand, contract or store energy. In all likelihood, there is no such thing as *nothing*.

Matter, it seems, is a uniquely integrated form of the vacuum, occupying a vastly smaller volume than the free vacuum from which it originated. By contrast, radiation bosons are a somewhat less compressed form of the ordinary vacuum, imbued - or able to be imbued - with energy. Huge volumes of space were compacted initially, if it is just "concentration" of the vacuum that now accounts for the huge energies seen associated with matter. Each nucleon, simply by discharging its graviton payload over billions of years, returns this same enormous individual volume back to space, in the form of an expanding universe. Matter itself seeks to revert to the ground-state form of the vacuum from which it originated, doing so quantum-mechanically by serial emission of gravitons. When every last graviton has been emitted, the universe will have lapsed back into the vacuum, and disappeared.

The existence of material particles is apparently the natural and only outcome of continued compression of the vacuum, since all atoms are the same. And all atoms are made of nucleons, which are made only of quarks. Once nucleons exist, a "clock" is started, and the laws of physics become evident and measurable. The nucleons produce gravity.

The number of material particles that must have appeared initially is in the order of 10^{83} (calculated as 10^{11} galaxies x 10^{12} stars per galaxy x 10^{57} atoms per star x 10^1 atomic particles per atom x 10^1 dark/visible ratio for galactic mass x 10^1 atoms originally produced for each one that still survives after the quasar era) - a colossal number. *Each* of these particles, moreover, had packed into it the equivalent of a kind of graviton universe, on the order of $\sqrt{10^{83}} \approx 10^{42}$ gravitons. Around 10^{125} gravitons, in total, were therefore assimilated in this early period.

Each of the 10^{83} neutrons surviving the quasar era can now, in principle, serially return about 10^{42} gravitons to the vacuum, and the return of each graviton to the vacuum can be thought of as a "tick" of the quantum clock. The time required to fully release all of these gravitons is clearly about 10^{42} quantum time-units. In terms of the Planck time - 10^{-23} seconds - this amounts to $\sim 10^{19}$ seconds, which is 10^{12} years. The current age of 10^{10} years therefore places the universe at about one hundredth of its maximum attainable age.

This creation model requires some enormous initial compression, driven by unknown forces. This quantum compression would be more like compressing a gas than a solid, so no over-dense region would exist which could define a "centre". The subsequent history of the universe would be that of a compressed quantum entity, relaxing back to its former state in a gas-like expansion at speed c, with no particular geometric centre. Gravity is the chief form of radiation in this relaxation, as *all* material objects are observed to respond to gravitational force, and the universe as a whole appears to expand.

Atoms drive material objects away from each other by continually emitting volume-displacing particles into the spaces between them. These particles repel each other slightly. Each emitted graviton represents a new incompressible quantum volume element in the vacuum. The vacuum expands in an attempt to accommodate the emergent gravitons at a constant number density, and this drives the general expansion of the universe. However, re-absorption of these gravitons by adjacent or line-of-sight particles also *removes* volume, causing material particles to deflect towards each other *as if* a force of attraction were also acting. As the scale lengths increase, the rate of this re-absorption drops away. Thus gravity attracts over short and medium distances, but at larger scales the rate of attraction falls away to reveal only the general expansion.

The recapture rate of gravitons is *ordinarily* much lower than the emission rate. Taking air as an example, the 10^{23} atoms occupying one litre of air under standard conditions occupy about 10^{-12} of that volume. Atoms in air are separated by about 10^4 times their own radii. At this separation, each atom receives from other atoms the equivalent of 10^{-5} square radians of flux per 4π steradians of receiving area, or less than 10^{-6} of its own outward flux. Any atom will therefore absorb - from a shell of neighbour atoms just 10^{-5} metres away - about 10^{-5} of its own flux output. Taken over 10^4 such concentric shells of radiating matter within the litre of air being considered, this means that about 10% as much is re-absorbed as is radiated. This is an upper bound, because many of these atoms stop the gravitons from more distance atoms from reaching the "central" atom, and the re-emission rate – presumably - is *not* driven up just because absorption is taking place[7].

If atomic emission is, on average, met by 10% re-absorption, the age of the universe will be extended by roughly 11%. For large portions of its age the universe also sees matter at near critical matter densities, and more so as it ages, as more and more matter finally gets located in the remnants of stars, where the matter density is near critical. Factoring in an estimate that 10% of the life of any material particle is spent in a near-critical radiation density, the result is that for 10% of its projected life an atom loses no *net* flux, lengthening its life by around 10%. The estimate of 1.1×10^{12} years moves up to about 1.2×10^{12} years, still not a significant change, although it would add 200 billion years to a universe otherwise headed for a 1000 billion year lifespan.

Some powerful immediate conclusions can be drawn. In this model, gravity would cause and sustain expansion of the universe, instead of halting it. The universe is not ultimately a geometric object, although can successfully be treated as one for most purposes. In reality, the expansion of the universe is a quantum-dynamical process. It started as a vacuum fluctuation in which 10^{83} neutrons systematically appeared in a concerted phase of creation, and - since each neutron independently returns around 10^{42} gravitons to the vacuum - the universe consequently has a life expectancy of approximately 10^{42} quantum time-units, which amounts to about 10^{19} seconds.

---

[7]   To this degree, this calculation of net gravitational flux differs from that of the flux equilibrium in a star.

Mass and Energy

September 24, 1235 AD: General Yoritsume and his army observed unidentified globes of light flying in erratic patterns in the night sky near Kyoto, Japan. The general's advisers told him not to worry - it was merely the wind causing the stars to sway.

September 12, 1271 AD: In Kamakura, Japan: The Buddhist priest Nichiren was to be beheaded, when an object appeared in the sky and the beheading was called off.

How does a change in the understanding of what causes gravity change our understanding of the meaning of *mass*? And what exactly is the link between matter and mass?

It is already common knowledge that the equation $E = mc^2$ suggests an equivalence between mass and energy. But it is a different relationship between *matter* and energy. Matter and mass are not the same. Under stationary conditions, mass will *represent* the amount of matter present, but with relativistic conditions one observes relativistic *mass* increase without a corresponding change in the amount of matter. Indeed, objects *foreshorten* under relativistic conditions, as if *less* matter were present. The *same* matter assumes a smaller and smaller size while the mass continues to grow, as energy[8] is pumped in.

There is a notion that mass can be "converted" into energy. But mass is a *measure* of something – not a substance – and as such cannot be "disassociated". Mass measures (to put it loosely) the "drag" of matter on space. Only when this drag is removed, by dematerialisation of matter, do large amounts of free energy suddenly appear. Matter dematerialises into gravitons, and the rest mass is now seen as having been an indirect measure of the amount of spin angular momentum (energy states contained within and around an atom) that is released back into the vacuum when the associated matter is suddenly converted back completely into its constituent parts.

During motion, an object's energy is raised above its rest mass energy, yet no new matter is in evidence. Mass increase (where gravitational mass gains inertial mass) measures how many new "photons" (excited graviton spin states) appear in association with an object that now moves. That is, mass measures total photon spin-density around an object, and has charge-like qualities. However, the action photons do *not* form part of the system called the *matter* of the moving object. Indeed, no experiment has shown a way to distinguish gravitational mass (mass associated with matter in a stationary state) from inertial mass (mass associated with the inertia of a moving object). As the action photons have no clock, they do not contribute to the matter – only to the mass.

This has a few peculiarities associated with it. A calculation made in the observer's frame will conclude that the *total* energy of a moving system has increased, because the number of aligned states in the vacuum surrounding the material has risen and because the object is therefore moving at a faster velocity than before. Another way to look at this is that the material's volume has diminished with speed, according to Lorentz transforms for length contraction, pushing up the number density of spinning photons per unit volume of space, within the volume occupied by the material. This is seen as a mass

---

8   Which, according to the earlier discussion, comes in the form of quanta of spin angular momentum.

increase. But from within the travelling frame, where the measurer is also foreshortened, no such increase in spin density is perceived. He consequently perceives no mass increase.

The internal measurer's clock also continued to operate normally. If energy were associated with matter in a travelling frame, or with the spin numbers of that matter, this change would also be observable from within that frame. Lorentz transforms would not work for outside observers, and angular momentum measured from different frames would not agree. Lorentz transforms only produce consistency when the extra energy is supplied only to the photons and gravitons associated with the motion, and not to the matter itself.

A boson, at best, sees a static and frozen universe. A photon does not interact with the universe outside itself. Yet experience reveals that this untouchable photon arrives in an observer's frame red shifted, even though its frequency should have been unaltered. Indeed it is accepted in all refraction experiments that *intrinsic* frequency of a photon is invariant. Red shift seems to indicate that some of the photon's energy - calculated from $E = hf$ – has disappeared in transit, if the new longer wavelength[9] is any guide. But this can only be if the value of h – Planck's constant - has altered. As this is not allowed, a paradox arises.

The options can be considered. Energy calculated from wavelength necessarily follows the equation $E = hc/\lambda$. If c is the *unchanging* rate of expansion of the universe, then it is the ratio h/λ that decreases, and does so in the same way that h must also decrease in the equation $E = hf$. This means that one should not *perceive* any change in λ at all, because the energy would re-scale on the basis of the change in h *alone*. Thus, the fact that a change *is* seen in λ seems rather to affirm the invariance of h.

If, instead, the value of h increases during the travel time of the photon (for "time" measured in the receiving frame), then *all* physical

---

[9] The explanation that the photon is a wave embedded in the metric of space, and that its wavelength-increase simply maps the expansion of space, is not applicable in the current model, as the photon sits *in* space but does not *propagate* through it or *stretch* (by virtue of attachment) with it.

dimensions will accordingly increase, as the quantum length unit uses this constant in its numerator. But the observer's diffraction pattern will then locate the wavelength associated with a photon of frequency *f* at the *same* wavelength it had in the emitting frame. This does not correspond to the reality that the photon arrives *red*-shifted[10]. The reality tends again rather to reaffirm the constancy of h.

With the constancy of h reaffirmed, an interpretation is – indeed - still needed as to why λ increases. The last possible solution seems to be to keep the product h*c* constant, while allowing *c* to increase with time (and h to correspondingly decrease). But this option has been disallowed already, by the earlier arguments that require h to remain constant. Therefore *c* must also remain constant. All of the arguments presented must therefore contain some fundamental *oversight*, if the paradox is to be resolved.

So, lets try something different. Let it be insisted, instead, that rest frame energy can *never* be lost from an isolated photon. Energy does not "leak" into the vacuum. It would then be necessary to reconcile what is actually measured by saying that frequency – which is a conserved quantity, along with energy – simply *measures* to be lower in the new frame, even if in reality it has never changed. And this can *only* be if the local clock has, in the meanwhile, sped up by the same degree as the apparent frequency shift. This would directly resolve the paradox, but it remains now to show how this could be.

All clocks are regulated by the presence of mass, and speed up as the universe ages. This makes sense, if the observation that the universe is continually expanding is accepted. Its matter density falls, relieving congestion of atoms. Yet it is for *this* reason – and not the Doppler effect - that one should argue that the red shift of light tells the observer that the universe is expanding. Red shift indicates to the observer how fast the clock was running when the photon was emitted. It expresses the difference in clock rates in terms of the difference between observed and rest wavelengths. It is *not* that h decreased with time, but that time sped up in the interim.

---

[10] The original rest-frame wavelength is known from identifiable spectral features like the H_α triplets

Because equivalence says that a strong gravitational effect is indistinguishable from that of a slow clock, the eventual red shift of the photon in its destination frame must always *look* as if the photon had escaped from a strong gravitational field. Indeed, the red shift must be equivalent to the effect of an integral of all of the gravitational force that has acted to expand the universe during the travel time of the photon.

$$E = MC^2$$

If clock rates change with the age of the universe, and if the energy associated with radiation appears to change with time when viewed on different clocks, then what is to be made of the famous relationship $E = mc^2$, and its dimensionality? Is this quantity truly constant?

Certainly, this elegant relationship accounts well for the lost rest mass in atomic fission or fusion. Atomic mass is lost and energy appears, in accordance with $E = mc^2$. But this, in itself, does not show how mass and energy link to clock rate, or vice versa. To achieve this link, it is necessary - first of all – to reformulate $E = mc^2$ as $E/c^3 = m/c$, which gives an energy density term on the left-hand side. The energy density E/c^3 is generally km/c, where k is a factor that allows "m" to be measured differently at different ages of the universe. As $1/c$ is given in units of seconds per metre, k/c becomes a quasi-linear time gradient in the universe at any given moment, which will here be denoted by Φ. That is, energy density is curiously defined by a gradient in time.

Since – at any age T of the universe - the rate of time appears to drop from $t = 1$ (maximum clock rate, standardised locally[11]) to $t = 0$ (where the clock appears to stop at the horizon), the observed change of clock

---

[11]   Choosing $k = 1$.

rate - per unit local time - is approximately $\Phi = \Delta t/T = (1 - 0)/T = 1/(R/c) = c/R$. In short, if $\rho_E = E/c^3$, then $\rho_E \approx m\Phi = mc/R$ - a quasi-linear relation[12], allowing reasonably direct interpretation.

From what was said earlier, "the average change of time rate per unit local time" (here, Φ) is also physically a way of quantifying the rate at which the atomic capture cross-section seen by gravitons increases with look-back time (or distance) through space, as this is responsible for the ongoing change in time rate. From the discussion about gravity, it was stated that the cross-section varies in direct proportion to the amount of matter present. So, dividing the energy density in a selected cube of space by this rate of change of cross-section gives a description of how energy density (and clock retardation) increases in proportion to the amount of matter present.

A second way to interpret $E = mc^2$ is as follows: the equation states that the gravitational effect of the mass is equivalent to that associated with an area-based energy density, E/c^2. It indicates that if all of the rest mass were turned back into energy, it would be represented by an energy quantity equal to c^2 times the mass in kilograms. Dimensionally, it must come out as an energy term. The potential energy[13] associated with gravity is $E = mgh + C$. Dimensionally, one has $E = mgh = kmv^2$ because $[mgh] = [kg][m/s^2][m] = [kg][m^2/s^2] = [mv^2]$. This is dimensionally similar to mc^2.

The introduction of the constant C – which must have units of energy - involves a change in the way potential energy should be understood, and its relationship to the energy of the vacuum. *Rest mass* is now equated with the non-zero potential energy of an object situated at rest in the vacuum, but with the observer's horizon representing the "zero" energy state at infinity. One can imagine a material object moving along this potential gradient, unimpeded by the gravitational attraction of all other material

---

12   One useful geometric tool that arises is as follows: since $k/c = \Phi = c/R$, then R – the current radius of the universe – is just c^2/k metres, and necessarily $k = c/t$, where t is the local time rate in standardised units. Looking back in time, one sees a clock running relatively slower (t smaller), as if mass were more massive (k larger).

13   The constant C accounts for the potential energy associated with the vacuum - usually set to zero, to simplify the equation into $E = mgh$.

objects. While travelling, it continues to release gravitons into the vacuum, emptying itself of "matter" and becoming steadily less massive[14].

It does not "fall" in a gravitational field to lose its energy. Rather, each atom expels its gravitons and relinquishes energy in this way. If loss of rest mass is not accounted for in this way, an object could arrive at the horizon with no remaining rest mass, and the problem of balancing the energy equation would there present itself. Any moving object in space should therefore be considered as slowly accelerating[15] along a "universal" gravitational gradient to the "edge" of space, all the while expending potential energy (and substance) until nothing is left.

"True" potential energy should – from this point of view - be defined as $E_{pot} = mc^2 \equiv mgh = m(c/T)(cT)$, the analogy suggesting that the ratio c/T corresponds to the horizon's own version of "g". Of course, c/T is also an *evolving* figure, as the "edge" of space towards which the mass accelerates is itself moving away with time. The dimensions of c/T are appropriately those of acceleration, so the universe as a whole can be imagined as providing this acceleration toward the horizon for any mass present in the expanding vacuum. One calculates this acceleration to be $3 \times 10^8 / 3 \times 10^{17} \approx 10^{-9}$ m/s^2 from any position, due to the cosmic expansion – a very small figure, but nowadays probably measurable.

However, Newton's Laws allow only *constant* motion, so the steady expansion of the universe implies that time must appear to alter in such a way that constant motion in the traveller's frame *appears* as acceleration in the observer's frame. The only way this can happen is for the rate of time to decrease with radial distance to the traveller, while the speed of light holds constant. But this is indeed what appears to happen from the point of view of a local observer[16].

---

14   It does not become "less massive" because it loses gravitons (which are, in any case, massless) but because the cross-section of the atom continues to diminish, reducing its gravitational interaction.

15   Although, locally, this acceleration is so small that it may still be interpreted as constant motion, and still give Newton's Law of Inertia.

16   This could be rationalized from a Newtonian point of view by arguing that the nearer one approaches to the horizon, the more the whole universe is trying to re-tard the motion, slowing the clocks of all observers to a standstill at the horizon.

To summarise this line of thought: the "mass" of an object measures the expansion force acting on an assembly of matter to drive it to the horizon of the universe, while the matter continues to expel its gravitons. This makes it an inertial quantity after all.

THE POTENTIAL ENERGY OF
THE UNIVERSE

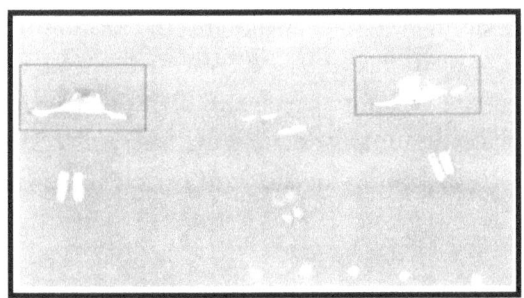

Figures 4 - 5: Two Cave Paintings from Tanzania. Both are estimated to be up to 29,000 years old. The one above is located in Itolo and depicts several disc-shaped objects, and some slab or cylindrical shapes. The painting below is from Kolo, and shows four entities (tall men, with variously oval or dog-shaped heads) surrounding a woman.

Most societies have ancient traditions of the gods taking human females and bearing children by them. The Bible's tradition has the "Sons of God" taking wives before the Noahic Flood, and their giant offspring wreaking havoc in the world, including fighting wars against each other. These giant offspring also went on to steal the wives of others for themselves.

To find a fascinating record of giant human remains all over the world, try a Google search under "giant human remains". Many human skeletons, male and female, in the height range 9-20 feet, have been unearthed, both in Europe and the Americas.

This new model appears to link clock rates and mass to the potential energy of the universe as a whole. It will be necessary, briefly, to characterize the nature of this universal potential energy field.

Everyday experience indicates that objects have energy apart from that related to their mass. An object is taken up a hill and then falls without any need of impetus, receiving kinetic energy. In that sense, this energy is usually forbidden to it and remains as a force of stress on the mass at all times, only relieved when the mass is finally freed to fall. The quantity of this stress is referred to as *potential* energy, as one understands it to be ordinarily largely unexpressed.

In the absence of local gravitational influences, this energy is not seen. Local gravity provides the acceleration that effectively converts E_p = mgh into E_k = $\frac{1}{2}mv^2$. As an object *accelerates,* its potential energy is converted into kinetic energy at an ever-increasing rate. In rearranging $mgh = \frac{1}{2}mv^2$, one arrives at the relationship $v = \sqrt{(2gh)}$. At the scale of the whole universe, with a force f acting to accelerate material toward the horizon, one would replace the final velocity v with c, and h (in mgh) with the distance $2cT$ to the observing horizon, where T is the current age of the universe.

Thus, $v = \sqrt{(2gh)}$ converts into $c = \sqrt{(4fcT)}$, making $c^2 = 4fcT$. And since the object starts with a speed of zero, and velocity grows linearly to c, only half of the distance cT is actually covered. The universe doubles in size, however, during the time taken to make this journey, and this further halves the acceleration rate. Thus, $c^2 = fcT$, or $c = fT$, where f is evidently 10^{-9} m/s^2, from considerations in the preceding section. Based on this, T (in seconds) would be equal to c/f, or 3×10^{17} seconds, which is 10^{10} years, as long as G is constant. This is the approximate age of the universe now, as required.

To treat the universe as a huge field of potential energy, a source and sink are needed for the field. The only place to put the cosmic zero potential is at the periphery of space, as judged from the observer's perspective. This means that every observer, regardless of when and where he is put inside the universe, will see the universal expansion as responding to the highest possible potential, c^2, locally, and zero potential at the horizon of space. The observer will perceive a downhill potential gradient of slope $-c^2/R$, which is a mathematically equivalent way to account for the expansion of the universe. The integral of this is a logarithmic function. The potential energy associated with gravity is just that provided by the expansion of the vacuum. Gravity may be *treated* as a static field, but actually it indirectly quantifies the evolution of the vacuum.

No other examples exist of a potential field where the gradient has its peak at *any* point inside the universe, but this is a quirk of General Relativity. In reality, gravitational force is not a potential field at all, in the normal sense, except when a mass is bound gravitationally to another mass in a locally static system. From this point of view, it is *not* entirely legitimate to think of potential for the non-static universe as a whole. Nevertheless, if one speaks of *zero* gravitational potential at the horizon (even if it is impossible to arrive there), it means that by the time one arrives at the horizon, the last graviton emitted locally is in the process of being absorbed back into another nucleon.

POTENTIAL ENERGY AND MATTER

In the Standard Model, it is the total self-gravitational potential energy of the universe that will ultimately determine whether it will "close" or not. The universe is "closed" if its own gravity can eventually halt the cosmic expansion and reverse the flow. And this boils down to whether there is sufficient mass in the universe to act as a gravitational brake to halt the expansion.

Going by the matter that is currently visible, and after adding in appropriate amounts of dark matter, there is an estimate available of the total self-gravitational braking force available to halt the expansion. But the model being promoted here will predict that all estimates of dark matter will eventually come in too low. This is because the issue of dark matter is irrelevant to the expansion of the universe, in this paradigm. This is asserted because atoms *continue* to produce gravitons, and continue to drive the expansion of the universe, at all times. The universe is not in free fall at all. Moreover, as they do this the atoms also slowly become less massive, further reducing the overall ability of the universe (on the Standard Model argument) to halt its own expansion by self-gravitation.

This peculiarity is reflected in recent attempts to reintroduce Einstein's discarded acceleration parameter in an effort to make sense of the measurements now being obtained. Cosmologists see no real evidence of a predicted slowdown in expansion rate now, compared to earlier epochs. In the Standard Model, no force of expansion continues to operate after the Big Bang, and the universe is in free fall. Its expansion should – at some level – show a slowdown[17]. Cosmologists are trying to quantify the rate of this slowdown. In the radiation-based model, the force of expansion, by contrast, continues to act at all times. With the constant introduction of gravitons into space, the universe has no chance to decelerate under its self-attraction.

The driving force is locked up in the atoms themselves. Atoms possess vast stores of potential energy, and as this is released back into the vacuum – in the form of volume-increasing gravitons – it is converted into the kinetic energy and impetus of the expanding universe. This is a gentle but ongoing and steady process.

It is necessary to extend pre-existing ideas about potential energy to see how this works. It is now necessary to integrate Newton's gravitational potential expression as an energy equation – giving $\int F_g \, dr = \int GMm/r^2 dr = -GMm/r + C = m\Phi$ as the potential energy associated with gravity. The quantity Φ was defined as the scalar potential energy associated with a mass, m, in the field of another mass, M. The integration constant is C, and its physical role is to add any *global* component of gravitational potential onto the local *zero*. C will not now be conventionally set to zero – otherwise all potential energies in the presence of matter will become negative quantities[18]. Instead, the energy of the universe will now be allowed to become zero only after the potential energy associated with its rest mass has also been accounted for.

Special Relativity assigns a test mass, m, a positive potential energy mc^2, when it is far from M and from all other masses – essentially at the local horizon of the universe. Inasmuch as this mc^2 is slowly relinquished back to the universe in the form of gravitons, it *is* potential energy, but is stored within atoms rather than in potential fields. If released suddenly, one sees just how much energy this really is. The only difference is that

---

[17] And have run into the problem that no such slowdown is clearly in evidence.

[18] This is tantamount to not allowing mass to be converted back into energy.

this particular form of potential energy is not immediately available. To account for this, it becomes necessary to set $C = mc^2$, and to rewrite Newton's integrated equation as $E = m\Phi = mc^2 - GMm/r$. This makes the per unit mass scalar potential $\Phi = c^2 - GM/r$.

Fermions therefore experience a potential of $\Phi = c^2 - GM/r \approx c^2$, when the distance r from all other masses (whose collective mass is M) is great. A fermion's *intrinsic* energy will remain as $E = m\Phi = mc^2$. For a typical fermion, the local gravitational potential energy term, $-GM/r$, is intrinsically small anyway – about 1.33×10^{-19}J at the Earth's surface - negligible compared to mc^2. It remains negligible even within Main Sequence stars. Only within the likes of a neutron star (at 10^{15} times the density of ordinary matter) is this finally a significant term.

The "gravitating" (i.e. atom-bound) mass in the universe obviously declines as the gravitons are released into the universe. The mass lost (mc^2) as radiation drives the continuing expansion of the universe, but also leads to a decrease in the total mass that gravitationally restrains the expansion. The amount of gravitating mass falls linearly with time, while the force of expansion stays constant. The total energy in the universe also remains constant. Between the two forces, the universe expands at a constant rate. This means it must be, and will stay, intrinsically "open".

When all of the atomic mass is finally dissipated, the universe will consist entirely of gravitons and photons moving in all directions at speed c. This mass-less universe will consist entirely of energy. This energy was all originally borrowed from whatever force it was that originally compressed the universe, to form the atoms. Without gravity, the free gravitons and photons will continue to move unimpeded through space, if no lower bounds upon the density of the vacuum exist. If a lower bound does exist, the photon universe will continue to expand until the density approaches this lower bound. The speed of light will fall away rapidly at that point, until even the photons and gravitons stop moving. The universe will then become completely dormant and truly invisible.

PARTICLES

Figure 6: Tall Men and UFOs
In this cave image, tall beings are depicted, and what looks like a UFO - with an ion exhaust trail - taking off in the background.

WHAT IS THE ATOM?

The foregoing discussion about gravity no doubt now raises a question in the reader's mind about what the atom really is. It is not some eternally unchanging object, itself fundamental to the universe, but is an evolving, decaying entity. How do the pieces of the atomic puzzle that we know about fit in with these new assertions about the source of gravity? How does the atom really work? Does it become more complex, or simpler?

Science is certainly an empirical business, and research is deeply concerned with number, ratio and solution set. But science is surely more than numbers – it is ultimately about concepts and principles. Numbers are just a way to "render" the underlying ideas. In this section, an attempt will be made to work back from the world of ideas to get some meaningful numbers. A model for the internal workings and world of the atom will be developed, based purely upon logical considerations about what can, or – as importantly – what *cannot*, sensibly be in the interior world of the atom. Ideally, experimental observations will confirm or refine whatever theory can be produced, further directing the thought experiment.

To begin with, it will be necessary to abandon all notions of electron or quark orbits as following rules resembling those describing gravitational orbits, as the discourse above has already pointed out that *mass* (upon which gravity acts) is only a way of measuring the atomic cross section available for graviton capture. Also, under relativistic motion, mass increases but matter shrinks. The matter gains nothing, but the matter-vacuum *system* becomes massive. Indeed, mass is more a property of the vacuum than of the atom. To the degree than an object is *gravitational* – i.e. attracts other objects – i.e. receives gravitons and causes space to become curved – it has an associated rest mass. And while this is a property for a particle, or for an atom as a whole as it interacts with incident gravitons, it has no real relevance at levels below the Compton radius, where different processes dominate, and the external world is essentially invisible. These internal processes appear to be highly ordered, generating gravitons on a regular basis and expelling them from the atom, whether the external conditions are relativistic or not.

In discussing the nature of the internal workings of the atom, little can be verified directly. One can only infer what is *in* an atom from what comes out when an atom is broken into pieces in the presence of an abundance of energy. Radioactive decay, fusion reactions and fission reactions provide clues, but it is otherwise impossible to break an atom into its constituent pieces without surrounding the whole process with massive supplies of what the atom itself contains so much of – energy. This makes things a little blurry. What investigators can, at least, glean from the observation of countless "smashed" atoms are rules about what can and cannot be produced, measurements of the energy and associated charges of emergent particles, and rules about the combinations of elementary particles that could account for what is observed. Atomic science is indeed entirely inferential.

Atom smashing has inferentially led to the theory of the quark, and the theory of quarks works too well to be wrong. This does not mean it is "right" either, but it is near the truth. The problem with quarks, though, is that one can never isolate one. They are not believed to have an existence beyond the atom. The quark is always an *inferred* object, with inferred properties. From this point of view, assignation of mass to a quark has dubious meaning. It means nothing more than the effect on the outside universe of any sub-atomic particle that contains quarks. If the mass of that particle is attributed entirely to its quarks, then the quarks are considered to carry that mass. But - in the current paradigm - this is a false conclusion.

The second tenet of any thought experiment about the atom is that due care needs to be exercised as regards the concept of "structure". Any structure we know about is made up of sub-structures, but in the atom it is ultimately necessary to get to those objects that are not themselves made of other parts. Having infinite hierarchies of particles and sub-particles, in which each type of particle is made up of sub-particles, only avoids the question of the ultimate particle, as well as being untrue to the quantum theory that space itself contains a smallest natural unit of size. In the end there can be only *non-material* "objects" with properties, combining with similar objects purely on the basis of those properties. This "symbol set" of nature may come down to no more than two or three basic elements.

Even apparently "dense" particles like neutrons are in reality mostly "hollow", containing nothing more than a couple of highly energetic quarks and supposed force-mediating gluons that account for most of the neutron mass. The quarks move about so much more rapidly than anything else in the external world, however, that an observer concludes the neutron must be "solid". Physicists have even gone so far as to describe atomic *forces* as nothing more than interacting force-carrying particles. Such particles "become" and "un-become", as forces are "statistically" transferred across the interiors of hadrons. But these forces are probably as much an illusion as the solidity of the particles they supposedly hold together. It is the quantum elements that will be the reality behind the *illusions* of mass, force and substance.

Probably the single characteristic that is identified as being shared by all of the smallest particles is the property of *intrinsic spin*. No known particle is spin-less[19]. Subatomic particles seem to interact with each

---

[19] The graviton, in the Standard Model, is either spin 0 (and thus spin-less) or spin 2, but in the current paradigm it is a spin zero entity. However, it is not considered to be a single particle, but rather the combination of two more basic, spin-½ particles.

other (attract, to varying degrees) through specific combinations of spin states. As a quantum quantity, spin is incapable of variation. It is not even incremented onto a particle. Each particle has a specific single spin state in which it exists. Particle collisions and theories show clearly that intrinsic spin is discrete and ambidextrously either left or right-handed. But spin is not the revolution of an object - or even of a field. It is even dubious to say that a particle "has" spin, as if this were an external property imposed upon the particle. Better to think of spin as becoming a particle. Spin is therefore at the heart of existence – an idea that seems at odds with the macroscopic experience that things do not naturally seek to spin.

Spin is certainly also linked somehow to energy, as energy differences are seen associated with aligned and anti-aligned spin states in electrons, as measured from line splitting in the interstellar 21cm hydrogen line. Scientists conclude that electrons can and sometimes do "flip" over and register a small potential energy shift. And electrons share spin and charge in common with quarks. Unlike the quark, however, the electron can experience the thermal world, and it's flipping has a real world statistical probability that is confirmed through the spectral fine structure of free hydrogen. It is unlikely, though, that particles operating deep within hadrons share any such "statistical" freedom. They are situated far below the Compton radius – the broad boundary separating the strictly quantum world from the external world where statistics apply.

Nature seems to be clearly bi-polar in its expression, and the universe seems to become increasingly binary as the journey into the atom continues[20]. Biology has male and female. Physics has left and right-handedness. Charge is positive or negative. Spin is up or down.

---

[20]   The question of north and south magnetic poles is subtler. One enlightening discovery of special relativity is that an oscillating relativistic electric field will be perceived as an oscillating magnetic field. Sub-atomic particles are both electric and relativistic, so - if electric particles prescribe loops, as they probably do - an oscillating magnetic field appears externally. Static electric charges certainly exist, but magnetic monopoles (a magnet with only a south or north pole – its field lines not closing on an opposite pole) have not yet been discovered, although have been predicted by some.

A third key tenet of this mental journey into the atom is that energy is not itself an elementary particle. It is, rather, a quantity or action of some kind that can be exchanged between all particles without prejudice. It is like the spiritual gift of the universe. Energy is somehow stored or represented in the rotational excitation states of vacuum particles. It is quantified in physics both in terms of rotational frequency and apparent photon wavelength. The sub-atomic state of course has no analogue of temperature, which is itself linked to heat – a measure of "free" energy. Only macroscopic processes can "freely" distribute energy in the way heat is spread around. Sub-atomically, energy is discrete and localized, and never in the form of heat. It is only expressed in the form of action.

The ultimate vacuum particle will also not "vibrate". Vibration – understood physically – involves the interchange of photons from atom to atom. There is little, in fact, that the ultimate elementary particle is really allowed to do, as so many of the macroscopic phenomena are disallowed at this very fundamental level. The mathematical reality is undoubtedly much simplified. No arguments from special relativity apply to the quantum world, and nor do the macroscopic experiences of time or gravity apply to those elements from which the quarks themselves are made.

Finally, the particle that emerges from the atom and becomes the vacuum particle must produce the expansion of space and the force of gravity. It seems eminently likely that this elementary particle, exhibiting all of these characteristics, will be the graviton itself. And the graviton must be intimately related to the quarks that constitute material reality.

Thus armed, we begin our journey into the atom.

WHAT IS THE QUARK?

The reality deep inside the atom is fundamentally different to that in the free vacuum. The atom has to act like a "pump", or "factory", for gravitons. When this pump stops, every physical process – even the integrity of the atom – ceases. The quark seems central to this pump, as it is the only particle available to facilitate this emission. The quark also becomes the central particle in the enquiries of particle physicists.

And yet one may be in danger of missing the point if thinking in this way. After all, if a quark is itself composed of gravitons, by what mechanism would it carry other gravitons out of the atom, while itself being forbidden to do so? It would need to transport gravitons in and out of the atom and simultaneously mediate the rapid interchange of forces between sub-atomic particles. This could only seem practical if gravitons were a natural by-product of the formation or collapse of force-transmitting particles, a view not touted here because the existence of force-transmitting particles is dismissed as an illusion.

It seems simpler, therefore, to imagine that the six types of quark experimentally "observed" to exist in the various atomic particles are just six separate energy states to which some elementary particle, or pair of particles, is elevated before finally – in a seventh energized state - being promoted beyond the hold of the hadron as a graviton. Because of nature's strong desire for spin pairing, the graviton that emerges is most probably the sum of two more fundamental units of existence, which will in this thought experiment be called *spirii* (singular, *spirus*[21]), because they are a form of spiral. Two spirii would form a quark, and since a graviton would also essentially be a quark, it is also a spirus pair.

The essential neutron would then be a Planck pile-sized object, with something like critical matter density at its heart – a miniscule sub-atomic entity from which spirus pairs continue to emerge in loops of action, displaying charge and carrying energy. At the completion of each loop, the pairs are promoted into ever more energetic loops, each loop showing a different associated charge and energy. Finally – after six of these loops - a pair has incremented enough energy to carry it beyond the Compton radius, to leave the hadron altogether.

At any one time, six excitation levels should coexist in the "central" (Compton radius > radius > Planck pile radius) neutron, with each loop one angular momentum increment behind the other in sequence. At each new increment – that is, at each "tick" of the clock - one pair escapes the neutron as a graviton. If the neutron or an atom is suddenly disrupted, it will yield up to six quarks at that moment, and a graviton. The pattern of loops is therefore as follows (refer to Figure 2, further down, for a full schematic representation of the idea; also take "quark" to mean "spirus pair"):

Loop 1 (quark #1) -----(a pair which has looped just one time)
Loop 2 (quark #2) ------- (a pair which has looped two times)
Loop 3 (quark #3) --------- (a pair which has looped three times)
Loop 4 (quark #4) ------------(a pair which has looped four times)
Loop 5 (quark #5) -------------- (a pair which has looped five times)
Loop 6 (quark #6) ---------------- (a pair which has looped six times)
Loop 7 (quark #7) -------------------(a pair which escapes at speed *c*)

---

[21] Avoiding the usual suffix "-on" (thus, *spiron*) for this primal entity, to set it apart as "unconstituted". The name suggests that intrinsic spin is the chief characteristic of these elementary objects.

Quarks 1 through 6 are within the neutron, but one pair always emerges as a graviton. The model suggests that each spirus pair (that is, quark) repeatedly loops back into the neutron centre – beyond the Planck radius – to pick up another energy increment. This would have to be an automatic, self-propagating, process. The quark encounters the Planck centre in a similar way to a wave encountering a wall: one or both parts of the quark suffer spin-axis or chiral changes. A chiral change would certainly emulate wave inversion at a boundary.

All six loops – at any one time – are incremented simultaneously. At each new cycle, the quark emerges with a new increment of angular momentum. Each orbit for a given quark may precess with each new cycle, as one quark pair repels another. The neutron would resemble a pulsating six-petal flower, whose unequal, gyrating petals collapse and re-emerge in cycles. Once the sixth petal collapses in any excitation series, it re-emerges just one more time, to escape the neutron for good. As this quark leaves, without re-engaging the nucleus, a new "petal" is born within the Planck pile to restart the cycle, to replace the quark that did not return. As the system gyrates, gravitons will be emitted equally in all directions, although in a strict sequence.

A rule can be suggested for how energy would need to be incremented at each loop. When a pair re-enters the nucleus (within the Planck radius) with one unit of energy, it will re-emerge with one more unit. If it enters with two units, it will leave with two more, and so on, following the rule that if it enters with n units of energy, it will collect n units more for the next loop[22]. This is the kind of fractal-like "boot-strapping" rule that seems necessary if nature is to maintain such

---

[22] At this stage, this rule is of course supposition: the rule might be unit increments at each cycle, or it could be that the increments at each loop follow the sequence 1, 3, 5, 7, …{odd numbers}, as the accumulating energy sums then produce square numbers. A third option is that a quark arrives with n units and accumulates $n+1$ at each encounter. The most "natural" system would be a single-increment system, or the system proposed here. However, if electron energies around an atom mirror the quark energies within, then the sums of quark energies should reproduce the $1/n_1^2 - 1/n_2^2$ energy structures seen for orbital electrons, and the odd number sequence might be preferred.

processes "automatically". In this way, after six loops, and having picked up energy a seventh time, an escaping quark would have acquired 1 + 2 + 4 + 8 + 16 + 32 + 64 = 127 units of energy.

With two chiral forms and two spin alignment states, and perhaps two charges each, up to eight possible combinations for any pair of spirii appear. These combinations would need to be considered collectively in any discussion about families of elementary particles. It should be mentioned that the Standard Model assigns various attributes to the quarks. Some have charges ±1/3, others have charges ±2/3, all have spin 1/2, and all have different rest masses. The description being offered here would need to reproduce these properties of the quarks. But while the proton's charge of +1 is built from three quarks having charges of +2/3, +2/3, and -1/3 in the Standard Model, and the electron's charge is −1, the current model may combine still more basic elementary charges to obtain the same overall results. For instance, it seems more sensible to suggest that every fundamental particle has a charge of either +1/6 or −1/6, so that any quark charges are just combinations of these. Pairs of spirii will then have charges of magnitude ($\pm 2n/6$) (n an integer), as indeed is observed.

At the macroscopic scale, similar charges resist being placed together, while dissimilar charges attract each other. But this is – indeed – macroscopic, and is a result of the effect of these charges on the intervening vacuum. At the level of a quark, this effect should not be seen, as vacuum curvature does not affect the quark. Indeed, it will probably be necessary to find an explanation for what electric charge *is* before it is possible to speak about the charge carried by a spirus pair. Possibly charge is an extrinsic, rather than intrinsic, quality – and would then almost certainly have to be related to intrinsic spin or chirality.

WHAT IS ELECTRIC CHARGE?

Electric force follows an inverse square rule, so charge – at the macroscopic scale – distributes itself over a surface and generates an "even" field around that surface. Research has identified the small charge carriers that distribute this charge over a surface as electrons, and these are considered to be point-entities. Each electron has it's own electric field, although one cannot say whether this quantity actually resides intrinsically *in* the electron, extrinsically *on* the electron, or externally *about* it. Certainly, electric *forces* exist in relation to static charges, but any force is just an illusion caused by the local curvature of space. Only *charged* particles appear to see this curvature. Uncharged particles seem impervious to it, although this may well be a cancellation effect – as any neutral particle contains equal numbers of oppositely charged particles within itself.

This charge-based curvature is generally understood to be quite apart from that caused by gravity, which has to do with distortion of the bulk flow of the vacuum around graviton-absorbing atoms. Also, the attractive *force* attributed to gravity is unidirectional, while electrostatic force is two-way. Unlike an atom, the vacuum does not "absorb" anything like a graviton, so no gravity-like (pseudo-massive) effect on the bulk flow should be seen from electric charge. The observed curvature created by electricity is therefore at first enigmatic.

Outside the hadron, similar charges always repel each other, but - from the Compton radius down - similar charges happily co-exist in spin-paired states, feeling no repulsion. This must speak loudly about the nature of electrostatic "force". Physicists use "selection rules" to get around these perceived logical anomalies. The Pauli Exclusion Principle is one such rule: it allows two electrons to share an electron orbital, so long as they are paired spin "up" and spin "down". Electron pairing (underpinning electron shell-filling theory) occurs – so the argument would have us believe - despite the apparent natural tendency of electrons to flee from one another, as spin-pairing generates some kind of bond that is stronger than the natural force of repulsion. But this certainly sounds like the argument of someone who refuses to accept the straightforward reality that electric force is simply not a feature of sub-atomic life.

The neutron has no charge at all, and the Standard Model sees this as meaning it contains one +2/3 and two −1/3 charged quarks. But when the neutron decomposes naturally into a proton and electron, the proton suddenly contains two +2/3 and one −1/3 quark, while the emitted electron has a net charge of −1. Clearly there is something a little awkward in this, as it requires a pair of quarks (carrying charges -2/3 and +2/3) to appear from nowhere[23] and disrupt the neutron, thereby precipitating a recombination of the quarks so that quarks with charges of −1/3 and −2/3 now combine to form an ejected electron with charge −1, with the remainder of the quarks combining to form the proton. Indeed, this theory actually suggests that an electron is some kind of merged quark pair. As well as the numbers work, the physics seems a little impractical.

It is significant that the quark that emerges as a graviton in the current model has no mass or charge, and moves at speed c. The electron, carrying its charge, *does* have a small mass and finds itself unable to attain the speed c. Spherically distributed electrical fields seem to interfere with the flow of the vacuum around a charge. A neutron, by

---

[23] Or else requires quarks to transmute.

contrast, has no charge yet *is* massive. In other words, mass cannot emulate charge, but charge can – to some degree – reproduce the effect of mass on the vacuum. Mass and charge seem to create curvature in the vacuum independently.

A neutron is not as "fundamental" as an electron, even if an electron can itself in principle be separated into two quarks. In the real world, neutrons regularly transform into protons. In any heavier atom, there is about a 60/40 ratio of neutrons to protons. It may seem odd that the proton and electron do not immediately recombine to form the more stable neutron, but this may be because of the need to somehow discard a surplus quark pair in re-forming the neutron. This recombination only happens again in the extreme situation of a neutron star, where electrons are "forced" back into protons, to form neutrons. So, the anomaly exists that unlike particles attract each other but do not ultimately merge back into a stable particle.

The other nuclear oddity - that the clustered positive charges within the nucleus do not fly apart due to electrostatic repulsion - is again usually dealt with by Pauli's Exclusion Principle (spin up/spin down pairings of protons), and secondly by defining a "strong" nuclear force – necessary to pull protons together with more force than the corresponding sum of electrostatic repulsions. But this is all somewhat unsophisticated. The explanation is probably as simple as this: like charges cause *atoms* to repel each other because they can distort the vacuum particles that separate those atoms, but within the nucleus of a single atom there are no such vacuum particles to affect. With no distortion (i.e. curvature) there is no apparent "force". So charges do not *feel* each other at all at sub-atomic distances. Electrostatic force is purely *macroscopic*, and the "strong force" is just a mathematical invention designed to overcome an imagined repulsion.

While charges can create "force", charges are not themselves forces. Quarks have charge, but little force arises within the atom in connection to these charges. What needs explaining in this whole situation is how a graviton – being a quark – invariably exists with no charge but emerges from an environment of charged particles. It can only be the case that at each of the six iterations of the quark within the atom, before its final expulsion, the quark acquires a different charge, which seems to arise from its cyclic re-engagement with the gravitons contained within the Planck pile.

The maximum charge on a quark has been established as $\pm 2/3$. This charge level may indicate the maximum charge that can be generated on a quark, where five states ($\pm 2/3$, $\pm 1/3$ and 0) are possible. It is possible to create a model in which charge is appended in a cyclic fashion with angular momentum, in accordance to the latest charge seen by the nucleus. That is, if the quark charge reaches $+2/3$ or $-2/3$, the nucleus reacts to reduce this excess. When a $+2/3$ charge reduces, it continues to do so until the charge finally becomes $-2/3$, and the increments then operate in the opposite direction. If the cycle could continue, it would just oscillate between these two charge extremes. The process would produce a zero-charge pair at the seventh cycle, if the starting charge for the pair were $+2/3$.

The process starts within the Planck pile, where each member of the original spirus pair gets an initial charge of $+1/6$ each (thus, the pair carries a total charge $+1/3$). This may be caused by the "impacts" of the quarks with the orderly arrays of spirii within the Planck pile (if this is seen as a "bounce"), at the ends of their loops. The quark sees one of its two constituent member particles suffer a chiral or spin inversion, and in the Planck pile one spirus suffers a simultaneous change to conserve momentum. This seems to set off the levitating and "looping" process that sees energy being incremented until eventually a pair are ejected from the atom altogether. This collision and spin reciprocation process somehow seems to account for the manifestation of "charge" in the first pair that emerges as "Loop 0". And – once charged – there seems to be no alternative but for the pair to prescribe loops within the atom.

The process sees spirus pairs loop ever farther from the graviton pile, with a changed chiral or spin state being accompanied by a charge change at each "bounce" from the nucleus. Schematically…

Loop "0"(inside the Planck pile): spirii spontaneously gain charge of +1/6 each (total +1/3).

Loop 1: spirii gain a charge of +1/6 each (total, +2/3 at first emergence in the "atom").

Loop 2: spirii gain a charge of –1/6 each (total, +1/3).

Loop 3: spirii gain a charge of –1/6 each (total, 0).

Loop 4: spirii gain a charge of –1/6 each (total, -1/3).

Loop 5: spirii gain a charge of –1/6 each (total, -2/3).

Loop 6: spirii gain a charge of +1/6 each (total, -1/3).

Loop 7: spirii gain a final charge of +1/6 each (total now 0).

At the seventh loop, the pair exits the neutron as a graviton, with no charge. With no net charge, the pair cannot create charge-based curvature in the external vacuum, and - with no ability to absorb other gravitons - effectively has no "mass" either[24]. As all seven loops are in the neutron at any one time, the sum of all the loops is always 1/3 (Loop 0) + 2/3 (Loop 1) + 1/3 (Loop 2) + 0 (Loop 3) + -1/3 (Loop 4) + -2/3 (Loop 5) + -1/3 (Loop 6) + 0 (Loop 7) = 0, the required net charge of the neutron.

It follows, of course, that the series (loop 0) can also arise with the first spirus pair being of charge –1/6 each (thus a total of –1/3). The cycle of charges then runs -1/3, -2/3 (loop 1, in the external atom), -1/3, 0, +1/3, +2/3, +1/3, and finally 0 at the point of emission. Thus, two types of neutron present themselves, which look the same in terms of overall charge and mass, but have slightly different internal stabilities.

In either form, if a perturbation causes one pair of loops charged –1/3 and –2/3 to merge and form an electron, the emission of this "over-dense" quark-pair will leave the remaining particle with a net charge of +1 – thus converting it into a proton[25]. For the set of cycles shown above, this necessary merger would take place between a set of loop 4 and 5, or loop 5 and 6 quarks; in the alternative particle configuration

---

[24] Indeed, by causing a small amount of local vacuum expansion at its appearance, it has a little "anti-mass", if this term can be used loosely.

[25] The Standard Model sees the proton as being a combination of +2/3, +2/3 and -1/3. This model sees it as +1/3, +2/3, +1/3, 0 and -1/3.

it would be a merger of loops 1 and 2, or loops 2 and 3, but a merger of loops 1 and 2 might be inhibited by its proximity to the Planck pile. From either version of the particle it is an electron that emerges.

It is significant that a charge of +1/6 (or, for the anti-particle, -1/6) initially arises, triggering the whole sequence. But from where does this randomly arise? Is it a "statistical" occurrence? With so few parameters of action available to these confined spirii, it seems that it must be something as simple as the change from ↑↑ to ↑↓ in the spirus pair that accounts for the arising of "charge". And this is tantamount to saying that charge is nothing more than a spin-stress. If handedness is included, eight combinations of ↑, ↓, ↑ and ↓ are made available – namely ↑↑, ↑ ↓, ↓↑, ↓↓, ↑↑, ↑↓, ↓↑ and ↓↓[26]. If each, in its surroundings, produces a different displacement volume in the vacuum, each could represent a different "charge".

Start as - ↑↓ (and ↑↓) are both 0,
        ↑↓ (and ↑↓) are both 1/3,
        ↓↓ (and ↓↓) are both 2/3,
        ↓↓ (and ↓↓) are both 1/3,
        ↓↑ (and ↓↑) are both 0,
        ↓↑ (and ↓↑) are both -1/3,
        ↑↑ (and ↑↑) are both -2/3,
        ↑↑ (and ↑↑) are both −1/3, and, emerge as
        ↑↓ (and ↑↓), which are both 0.

The reverse sequence gives the alternative neutron form.

These eight possibilities (and their conjugates) would correspond exactly to the eight loops of the graviton (Loops 0 through 7). The state changes described above follow the repeating sequence (where S1 denotes spirus 1 and S2 denotes spirus 2): "change handedness of S1, flip spin in S2, change handedness of S2, flip spin in S1". Minimum tension (i.e. *charge*) occurs when the two paired spirii share the same handedness and opposite spins, and maximum tension occurs for

---

[26] Note: ↑↑, ↑↓, ↓↑ and ↓↓ are indistinguishable from the first four combinations.

same spins and opposite handedness. The meaning of + and − is not numerical; indeed, the signs are arbitrary. While ↓↓ (and ↓↓) are both 2/3, and ↑↑ (and ↑↑) are both -2/3, all four have magnitude 2/3. But two combinations will cause the neighbouring pairs to move slightly apart, while two drive them slightly closer together.

Figure 7: Two-dimensional schematic of the atomic cycle.

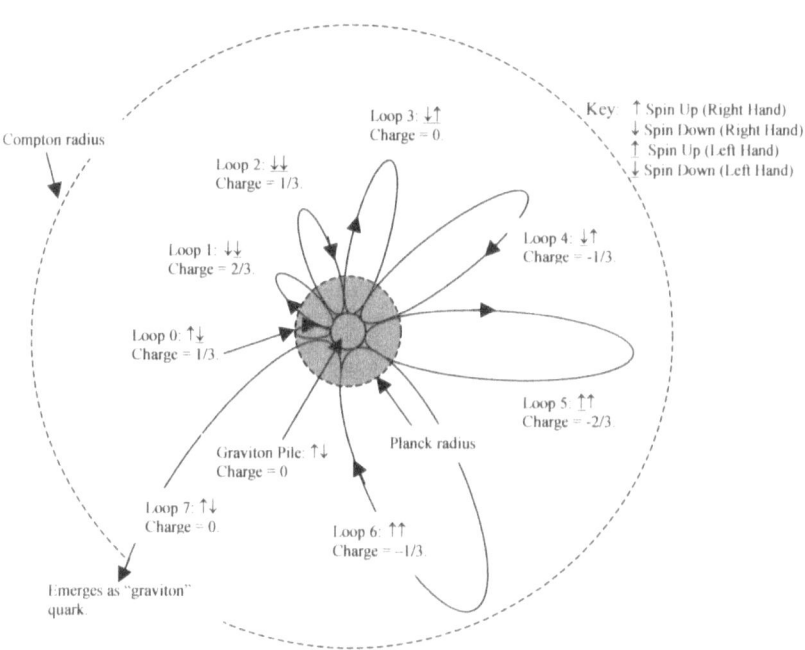

WHAT IS THE ELECTRON?

The electron combines a −2/3 and −1/3 quark. The eight possible arrangements within the electron must keep the components of the −2/3 and −1/3 from rearranging themselves, or the charge on the electron would alter. If the eight combinations are restricted to pairs of particle lobes, then four combinations present themselves:

Any lobe: ↑↑ + ↓↑, ↑↑ + ↓↑, ↑↑ + ↑↑, ↑↑ + ↑↑, ↑↑ + ↓↑, ↑↑ + ↓↑, ↑↑ + ↑↑, ↑↑ + ↑↑.

Same particle lobes: ↑↑ + ↓↑, ↑↑ + ↑↑, ↑↑ + ↑↑, or ↑↑ + ↓↑.

Figure 8: Schematics for the Electron

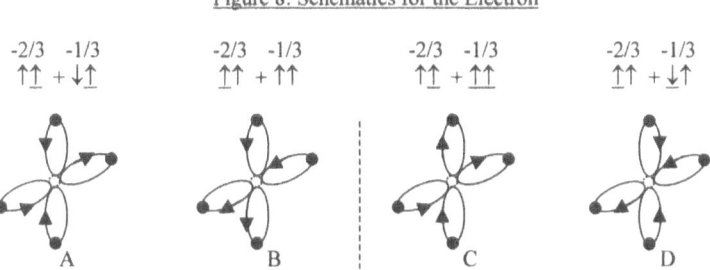

Spin-spin electron pairings can occur between A and B, or C and D, and similar pairings for the four anti-electron configurations.

The fact that the vacuum responds to charge when it is present shows that charge "induces" some kind of change in the vacuum density. But what else is there to alter in the vacuum except the spin state of another spirus, or the separation of the spirus components in a graviton? If vacuum states can "expand" under this influence, it is equivalent to creating "curvature" (a local bulge) in space, and this is measured as a *force*. A "positive" (the assignation is arbitrary) charge may increase the vacuum density, while a negative charge decreases it. One swells space, while the other shrinks it a little. Fields of force are felt. Even a photon or graviton will feel this bulge, and electric fields *do* indeed deflect light beams, and therefore gravitons.

If nature had not come with electric charge – and with the utility of the mobile electron - it would have been unable to produce photons. The universe would have been quite static. When electrons orbit around atoms, and jump between excitation states, they absorb or release energy. Electrons in different energy states become the medium of chemical bonding, physical chemistry and life itself. The interchange of photons between atoms transfers heat and light from stars to planets. Radioactive heat produces the molten cores of planets, and from convective cores come the planetary magnetic fields that protect life. Thus, electrons and photons both create and protect life.

WHAT IS THE PHOTON?

Armed with all of the foregoing, the final task in setting forth our model will be to produce a consistent description of the photon and of its relationship to the graviton. As the photon – in this theory – becomes the only *necessary* boson in the universe, it is probably the only boson that actually exists. Others – like the mediating bosons of the "weak" interactions between quarks and electrons, and quarks and neutrinos - have been invented to account for electrostatic forces between protons and electrons, but presume that electric force is *felt* within the Compton radius. The current theory rejects this belief. (Even in the Standard Model, these bosons seem to present a mathematical problem, because they violate parity constraints[27].)

A description is needed, however, of how the photon arises, transports itself and finally meets its end. When a photon "springs" into existence, it is observed to travel instantaneously at the speed of light. It finally delivers its payload of energy to some atom and "disappears". It seems

---

[27]   Not that one should be too dismissive of the Standard Model. All fractal systems of the Mandelbrot type, for instance, can be produced in pairs that are non-identical mirror images of each other (thus parity violating). Fractals – which are complex number entities of arbitrary dimension - *could* perhaps form a model for the internal structure of particles in the Standard Model.

more than likely, therefore, that the photon is not a particle at all, but simply an energized version of the graviton. When the graviton de-energizes (by unloading the energy that turned it into a photon), the photon seems to "disappear". However, the graviton continues to exist invisibly.

This model has the beauty of simplicity. It is necessary only to show how the graviton becomes energised, and to discuss the allowable states of such excitation. The graviton is an entity in which two spirii with the same spin-handedness rotate about each other with opposed spin axis vectors. The graviton does not "propagate" at all. It displaces the vacuum about it at the speed of light, joining a sea of other gravitons doing exactly the same thing. As the graviton leaves the atom, an electron may at that moment be attempting to make a downward transition between orbital shells. It cannot do so, however, unless energy is transferred to the emerging graviton. Electrons and gravitons therefore coordinate energy transfer between themselves, in a way that even photons cannot do among themselves.

Similarly, an electron cannot be re-excited without an incoming graviton (in the form of a photon) supplying it with the requisite amount of energy. The discrete energy chunks exchanged are in fact increments of angular momentum. Angular momentum is virtually the basic quantum entity in the universe – its *means of exchange*, as it were - and is the *form* in which energy is stored in the vacuum. It is also a well-known form in which energy can be stored in macroscopic systems.

Just as linear momentum is an indirect measure of energy for an object travelling in a line, so also angular momentum indirectly measures energy for an object that is rotating. It is not – in point of fact – energy that is conserved in the universe, but total angular momentum. And a graviton, albeit exhibiting no mass, is nevertheless an entity that can rotate. The permissible orbital (and, thus, excitation) states of an electron are not determined by anything but increments of angular momentum, which amount to increments of rotational energy.

Given that the spin radius of the graviton is fixed, and also the spin rate of the spirii within the graviton, it is only possible after this to spin the entire graviton (i.e. the whole *system* defined by the spinning spirii). The angular momentum quanta produce graviton rotation, and increments of rotational energy lead to increments in the rotation rate. In the quantum world, if the first quantum produces a rotation rate R, the next quantum moves that up to a rate of $2R$, without radius increase. A photon is thus a rotating graviton. The rotational energy is both added and removed by increments.

Photon frequency is therefore just the spin rate of the graviton, and the frequency is a linear function of the energy, and vice versa. Although theorists assign a wavelength to the photon, this is just an illusion. But the illusion indicates, at least, that the graviton spins around an axis which points in the direction perpendicular to its apparent propagation through space. In the diagram below, the "spin" is actually out of and into the page. The photon, in reality, is neither wave nor spiral; it is nothing more than rotational energy stored in the rotation state of a graviton.

Figure 9: The Photon as Rotating Graviton

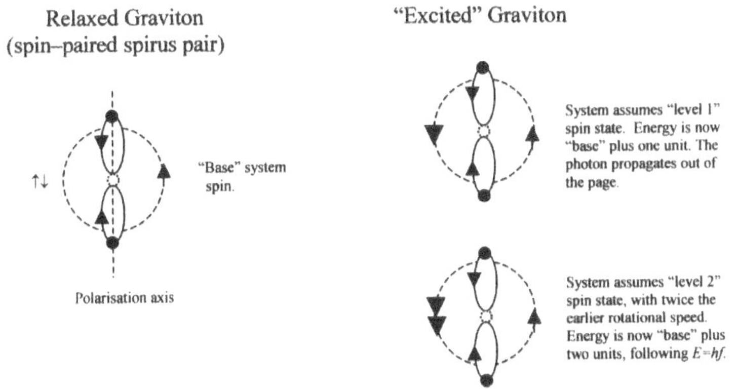

Relaxed Graviton
(spin–paired spirus pair)

"Base" system spin.

Polarisation axis

"Excited" Graviton

System assumes "level 1" spin state. Energy is now "base" plus one unit. The photon propagates out of the page.

System assumes "level 2" spin state, with twice the earlier rotational speed. Energy is now "base" plus two units, following $E=hf$.

Photons – and probably gravitons also – respond to an external magnetic field. The plane of the photon's associated "electric vector" is deflected in a magnetic field – a phenomenon known as Faraday rotation. The beam's axis is rotated in proportion to the strength of the magnetic field passing through that plane. In Figure 5, the polarisation axis of the photon is vertical, and the photon propagates out of the page. A magnetic field from left to right causes this axis to rotate left or right.

Figure 10: Polarisation of the Photon

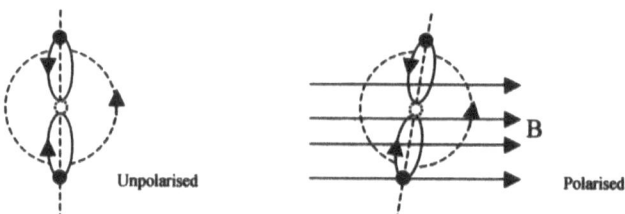

LINE SERIES

The photons seen in the line series for hydrogen are well accounted for by the Bohr model of the hydrogen atom. This seems to get "right in principle" the energy structure of the electron shells in the hydrogen atom (i.e. $E \propto 1/n_1^2 - 1/n_2^2$, for principal quantum numbers, n_j). Electron orbital (potential) energy goes as energy density; that is, it falls as $1/r^2$ around a point source, namely around the atom. Photon energies are just differences in these $1/r^2$ energies, where the values of r are the average orbital radii of the electrons. The inverse-square rule, and the relationship of the differences between electric potential energies (minus ground state energy) at different orbital radii to the energy of emergent photons, is clear. While this might have seemed a stroke of genius by the authors of the Bohr model, it was in a way also suggested by an equation between energy and work, and Einstein's equation, $E = mc^2$. An electron orbiting under the influence of electric force has electrostatic centripetal acceleration of $F_e = mv^2/r$. Replacing v with c, this becomes $F_e \approx mc^2/r$. Now since the energy consumed as work is calculated, in general, as $E = \int F.dr$, and since electric potential energy around an atom is calculated with F_e as a fixed static force, we have $E = \int F_e \, dr = F_e \int dr = F_e r = mc^2$. The energy of photons is therefore linked to the electrostatic potential of the electrons that will produce that photon.

Electron orbital energies probably stem directly from the orbital energies of the quarks within the Compton sphere. So ultimately it must be the quantum states of these quarks that fix the quantum states for the orbiting electrons, and thus the energy differences between these states. This then determines the available photon energies. As 99% of the universe is hydrogen, which contains just one nucleon (originally the neutron, but – with sufficient time now for radioactive decay - more commonly a proton), the majority of photons in the universe represent transitions of the hydrogen series (e.g. Lyman, Balmer and higher order series). Of course, where *many* nucleons populate the nucleus (A > 1) the mechanism for graviton emission, and the combinations of the available electronic states, becomes steadily more complicated. These atoms are responsible for all of the other spectral lines (i.e. photons) seen in nature, but all of the same principles apply.

Graviton and Photon Absorption

The model for gravity works if atoms can absorb as well as emit gravitons. Gravitons are charge-less, mass-less particles, with their spirii in either the $\uparrow\downarrow$ or $\underline{\uparrow\downarrow}$ arrangements. It is tempting to think that the graviton interacts with an electron and in that interaction manages to transfer the momentum, but no simple mechanism suggests itself. The interaction would involve three interacting quarks, which seems a little too specific to be a general process.

More probably, the photon or graviton is directed straight to the Planck pile, to be reabsorbed there. The photon's linear momentum is removed on the way in by depressing the orbits of the quarks slightly – measured externally as the "gravitational" effect. In the case of the photon, once it reaches the Planck pile, its angular momentum is distributed back over one or more of the existing quark orbits. One of the non-$\uparrow\downarrow$ quarks, in receiving some of this angular momentum, could find itself suddenly forced to orbit beyond the Compton radius. Once found "outside" the atom, a charged quark would probably seek to discharge its excess energy onto an electron, so recovering its own orbit and returning to balance with the other quarks. The electron would take this energy and find itself promoted into an "excited" state. As the electron made an upward electronic transition, the quark would drop back to its proper orbit, and the photon would seem to have disappeared.

How Do Photons Superpose?

Even in an expanding universe, where gravitons and photons are all in "free fall" in the general expansion, it is possible for them to meet and "superpose". When laser beams are passed through each other, the photons are presumed to superpose in the same fashion as waves, and individual waves then carry on after the superposition. But there is actually a catch in this understanding of overlapping waves, and that is the necessity that the energy sum of the over-lapping quantum states is *also* an eligible quantum state.

This is not a problem if the energy states of all incident photons are simply incremented sums of the basic angular momentum quantum. Any two or more states, when and if they are added, will just determine another acceptable state, and the vacuum will afterwards redistribute the quanta according to the associated incident momentum of each photon (proportional to each photon's frequency), thus reconstituting each wave exactly. There is nothing analogous here to "heat" loss in the vacuum. The interaction will also not assume the characteristics of some elastic collision of solid objects.

Yet, in all this, there is the inescapable problem that the miniscule spirons – which are locked in spin-pairs to form the gravitons that carry all of these quanta of angular momentum – can probably not occupy the same points in space[28], and cannot "merge" like fluid waves. Space is not ultimately fluid. They must then find a way to pass each other non-destructively.

At some level, gravitons must "push" against each other, because graviton density in the vacuum evolves in an attempt to hold to a constant value, even as gravitons continue to be introduced from every atom. Every emergent graviton occupies a specific quantum volume in the vacuum. The integrity of this quantum volume acts like a "selection rule", disallowing two gravitons in the vacuum beyond the atom to simultaneously occupy a volume less than, for instance, the Planck radius. So, if two gravitons look like they are going to impinge on the same space, the vacuum will need to expand locally to allow both to pass at a "non-conflicting" distance.

The Planck radius, however, is a truly miniscule distance – far smaller than the average perceived size of photons. Photon beams would need to be truly intense to push the number densities of photons to the limits where the average separation was smaller than the Planck radius. In such an extreme situation, where gravitons cannot easily find a path past each other, the colliding photons begin to assimilate into matter, and a new atom can be born. This has been reproduced in particle accelerators already.

Gravitons swerving around each other and causing small local expansions is undoubtedly part of what – as a cumulative effect - drives the expansion of the vacuum. The quantum vacuum speed is also sufficiently high (at least 10^5 times the graviton speed) that information about necessary density adjustments is transmitted around an impending contact zone in sufficient time that all incident beams can make adjustments and negotiate the "contact" zone without contact. Overall, any incident beam will continue in what is macroscopically determined to be a straight path.

---

[28]   Else they would no doubt do this in the Planck pile, before their expulsion. However, perhaps their incident speeds could have something to do with changing the properties of their interaction in the free vacuum.

SUMMARY

The discussion in the preceding sections can be summarized in the following statements about the spirus:

- It exists in left-handed and right-handed forms

- The left- and right-handed forms both have intrinsic spin, and never exist without this

- A pair can exist in either parallel or anti-parallel orientation

- The spirus knows nothing of time or physical dimension

- It can combine with another particle like itself, in a spin pair, to produce a "quark". Quarks are characterised as follows:

  o Quarks are responsible for electric charge: charge measures the "stress" generated by different combinations of handedness and spin orientation of the constituent spirii

  o The basic quark charges (±1/3, ±2/3 or 0) are combinations of a still more basic unit "charge", equal to ±1/6 of the proton and electron charges

o Charged quarks do not feel any electrostatic forces – these are purely macroscopic effects; macroscopically, however, charge can change the "density" of the vacuum, appearing there as a stress

o Quarks are responsible for the available energy states of "orbital" electrons

o Quarks can be energized by quanta of angular momentum

o The synchronised pulsations of six spirus pairs – each pulsation staggered from the next by one excitation state – creates the entity known as the hadron; each spirus pair is a quark

o The atom regularly releases an energised ("seventh") spirus pair as a "graviton". Gravitons are characterized as follows:

   ▪ Gravitons are a form of quark (the hadron's unseen seventh quark)

   ▪ Gravitons have no rest mass or net charge

   ▪ Gravitons can rotate, and a rotating graviton is called a "photon"

   ▪ The graviton carries energy in the form of rotational energy, by storing quanta of angular momentum (turning it into a photon)

   ▪ Gravitons (as photons) transport energy in the universe

      • The universe cannot exist without transfer of energy in the form of photons

      • Polarisation of light shows that photons react to magnetism as if they were "charge-like" objects; although the net charge on a photon is the same as that on a graviton – zero – the two half loops of the photon look like small equal but opposite current loops

- No two gravitons can occupy the same space in the vacuum; gravitons and photons thus tend to repel one another as they approach

- Graviton repulsions drive the expansion of the universe

- Re-absorption of gravitons produces the attractive force of "gravity"

- Graviton re-absorption is responsible for distortion in the bulk flow of the expanding vacuum, and so "curves" space.

- The spirus can combine with other particles like itself, to produce an electron

  o An electron is the combination of two quarks; there are four such combinations for the electron, and four for the anti-electron

  o An electron is produced when a neutron radioactively decays into a proton, and excess quarks (one of charge −1/3 and the other of charge −2/3) need to be ejected from the neutron

  o Electrons can exist together in anti-parallel spin pairings

  o Electrostatic force operates only at distances well above the electron-electron bonding distance

  o The existence of electrons is the basis for all life

.

A Creation Scenario

12,000 BC China: The Dropa, the name given by the Chinese ancients to visitors from Sirius, came down from the clouds with their air gliders.

3000 BC China: According to the book *Memories of the Sovereigns and the Kings*, published in the 3rd century AD in China, in the third millennium BC, before the birth of Huang Ti or of Chi You, "sons from the sky", would descend to Earth on a star which was the shape of a saucer.

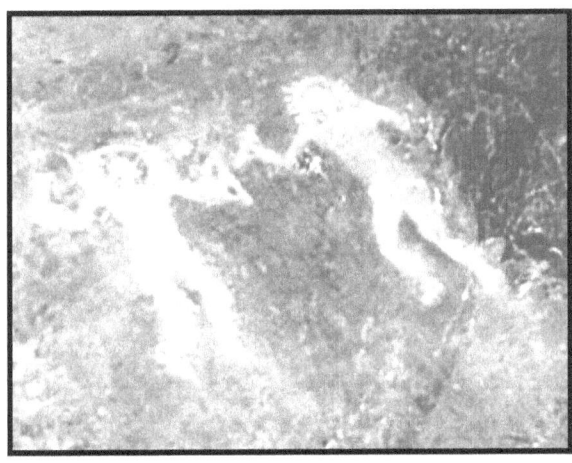

Figure 11:
Astronauts on
Rock
This rock
painting dates
from *c.*10000 BC,
and is found at
Val Camonica,
Italy. It seems
to depict two
godlike beings
with radiating
faces, holding
what could
be scientific
implements, or else two warriors wearing helmets bearing clubs and small shields (as they seem to be fighting).

In the beginning, something caused the vacuum of space to begin a dramatic collapse. This continued compression saw the volume density of elementary vacuum particles steadily rise. These elementary particles – called spirons – began to arrange themselves in an orderly fashion, as the continuing compression began to impose the need for efficient spatial organization. At still higher densities, spin-pairing of spirons began to take place, again to lower stress. Eventually, a critical density was reached, where the spirons could be forced together no further. Highly ordered, densely packed, spin-paired spirons began to evaporate as quickly as they were being assimilated. At this moment, the material universe was born.

The vacuum compression phase began to slow exponentially as matter began to precipitate from the vacuum in the form of neutrons. At the beginning of the slowdown these particles began to form at prodigious rates. As the compression rate eased off, so did the rate of new neutron production. But compression and precipitation did not fully abate until at least ten times more than all of the matter now seen in the universe had been produced.

During this early assimilation phase, there could have been only insignificant local expansion in the universe, even if some neutrons were already radiating for some of the time. The growing cumulative radiation of many new neutrons eventually began to force back the compression wave and to counter the influx of new spirons. By the time a balance point had been reached, so that the expansion of the universe could begin, some 10^{83} neutrons had already been formed, and the production of new atoms was in exponential decline. The universe now began its "rebound", and the expansion rate climbed steadily to c.

For the radiation in the expanding universe to remain at constant surface density, it would have been necessary for n^2 neutrons to be produced in every n time-quanta[29]. The point at which this particular

---

[29] This relates to that period in which neutrons were still forming and had begun to radiate gravitons, so that their collective radiation pressure had become sufficient to overcome the compression wave. This phase was, of course, temporary, and the compression phase has long since finished, as has the neutron production phase. It was not long after the first neutrons were formed that the radiation backpressure began, and continued to be sufficient to both

balance occurred can perhaps be regarded as the point at which the universe began, if the prevalence of expansion over compression is taken as denoting the point at which the universe is born. If each of these n^2 neutrons released a single graviton in every quantum time unit, there would be always be n^2 gravitons at the nth tick of the quantum clock. At all ages nt (age = n quantum time units, each of length t) the n^2 neutrons would each have contributed n gravitons to this expanding space. The expanding space allowed $(n + 1)^2$ new gravitons to be created at the *next* tick of the quantum clock. Since, at age nt, all n^2 particles created would each have expelled at least n gravitons into the vacuum, this vacuum will always contain at least n^3 gravitons at time t. This would facilitate a constant-density expansion.

Under these specific circumstances, neutron numbers would have built up according to the series sum $N(t) = n^2$ particles at age nt, necessarily following the simple sum $N(t) = 1 + 3 + 5 + 7 + 9 + ...$, which is just a series sum of the odd numbers, giving $N(t) = n(n -1)/2$. At any age nt, the vacuum fluctuation – i.e. the universe, in which time has now sprung into existence - would already have reached a volume of $4\pi(nr)^3/3$, where r is approximately the Compton length h/mc, if using the Planck time as the counting measure. The energy density will have followed

$$E_{total}/(4\pi(nr)^3/3) = E_{total}/(4\pi(nh/mc)^3/3) = E_{total} (3m^3c^3/ 4n^3h^3\pi) = 3E_{total}(mc/nh)^3/4\pi$$
$$= 1.27 \times 10^{44} E_{total} /n^3,$$

(using current values of h = 6.625×10^{-34}, $c = 3 \times 10^8$ and neutron mass of 2×10^{-27} kilograms as a modern scale comparison of this early universe). As the n^3 vacuum particles each occupied a volume of $4\pi(h/mc)^3/3$, the size of one of these vacuum quanta is about 10^{-44} m^3, about the current volume of the neutron.

---

drive the expansion of the new universe and to halt the compression flow. The expansion rate must have built up steadily to the speed c, reaching this speed as the last new neutrons were being formed. However, before reaching this speed, the universe must have reached a point where its vacuum density failed to continue under conditions of constant density. The constant density situation just marks out the point where the force of expansion finally exceeds the force of compression.

If the n^2 neutrons per n time-quanta principle were an average figure for neutron production rate in the early universe, it would have taken about $10^{41.5}$ time-quanta to form the 10^{83} neutrons. This is some hundred billion years. Clearly the production rate must have been far in excess of this in the earliest phases. This phase must have come to an end after about $10^{39.5}$ time-quanta, or one billion years. This suggests average graviton assimilation rates some ten thousand times greater than the current graviton release rate, and peak assimilation rates may have been as much as a million times those seen now during the relaxation phase. The universe in which the quanta are now systematically returned to the vacuum should endure for ~10^{42} quantum time units, where a quantum of space is returned from every nucleon at every quantum tick. When each of these 10^{83} nucleons is finally exhausted of its 10^{42} gravitons, the universe will have expired.

The time quantum for counting is the period h/mc^2 – the time taken to cross the quantum radius at the speed of light, c. Because gravitons are *re-absorbed*, as well as being emitted, the ageing process is slightly extended, as if pushing up N. The other thing that can affect this aging is the evolution of matter density in the universe.

To establish the magnitude of these effects, a calculation will be attempted. The situation of critical flux density will be taken as being that at which the quasars became active, somewhere near 10^{17} seconds. Net flux will be determined by merging the effect of falling matter density with the effect of a shrinking atomic capture cross-section on graviton re-absorption.

To illustrate the principle of the calculation, imagine an atom radiating a constant ten units of flux, set within a matter distribution that returns an inward flux that falls with the age of the universe from ten units to zero. At some point in the atom's life, the inward flux will be three units, and the net outward flux should be seven units. But if, in the meanwhile, the surface area of the radiating atom has also shrunk to seventy percent of its original value, the recapture rate will not be 3 units, but 2.1. The net flux will not therefore be 7 units, but 7.9.

The optimal outward flux from an atom is 10^{23} gravitons per second, so $F_{net} = 10^{23} - F_{in}R$, where R is the recapture proportion. The capture cross-section will be described in normalised units from 1 to 0 in the post-quasar life of the universe (spanning from 10^{17} to 10^{19} seconds, bearing in mind that 10^{17} is just 1% of 10^{19}). The starting surface area of the atom at 10^{17} seconds is about 1.26×10^{-29} m².

As the atomic cross-section decreases with time, the graviton re-absorption rate falls. The atom becomes an increasingly small "target". To model this, the Compton volume of the neutron is taken to be about $10^{-44.5}$ m³. There are 10^{42} gravitons per neutron, so each graviton accounts for about $10^{-86.5}$ m³. With 10^{23} gravitons emitted per second, the volume lost is $10^{-63.5}$ m³/s. The time derivative is $dV/dt = 4\pi r_a(t)^2 dr_a(t)/dt$, and this quantity is the constant value $-10^{-63.5}$ m³/s. It is notable that the per-unit-mass measure of how dV/dt changes with time has units $[m^3/s^2]$ – the same units as those of the universal gravitational constant, G.

The result is that the rate of change of atomic radius, $dr_a(t)/dt$, is given by

$$dr_a(t)/dt = (-10^{-63.5}/4\pi)(1/r_a(t)^2) \text{ m/s}$$

describing a 2nd-order hyperbolic slowdown in radius decrease, reflecting a constant rate of volume decrease.

The rate of change of surface area – i.e. the all-important capture area – is

$$d(SA)/dt = d(4\pi r_a(t)^2)/dt = 8\pi r_a(t)dr_a(t)/dt,$$

or

$$d(SA)/dt = (-10^{-63.5}/4\pi)(8\pi r_a(t)/r_a(t)^2)$$

$$= -2 \times 10^{-63.5}/r_a(t) \text{ m}^2/s.$$

The decrease in capture area also falls hyperbolically, but in the 1st-order with time.

Having set the atomic radius at $r_a(t) \approx 10^{-14.8}$ m at the *current* era, this has given an equation for the evolution of surface area from *now* on. To correct this for the Compton radius applicable at the Quasar Era, it is necessary to substitute $t = -5 \times 10^{17}$ seconds into the integrated version of the equation for dr_a/dt. This shows that the radius at creation was 5% $((10^{-44.5} + 5 \times 10^{-46.5})/10^{-44.5})$ greater than now. Incorporating this into the cross-section formula, to place it at the same place along the time axis as the emission equation, requires an amendment to

$$d(SA)/dt = -2.1 \times 10^{-63.5}/r_a(t) \text{ m}^2/\text{s}.$$

To make this parametric in t, the relation[30]

$$r_a(t) = \sqrt[3]{3 \times (10^{-44.5} - 2.1 \times 10^{-63.822} t) \Big/ 4\pi}$$

must be fed in, derived later from arguments presented in the section *When the Compton Radius Reaches the Planck Radius*. Scaled as a proportion of the atomic radius $r_0 = 9.1 \times 10^{-16}$ meters, this gives the *relative* radius as a function of time as

$$r_a(t) = \frac{\sqrt[3]{3 \times (10^{-44.5} - 2.1 \times 10^{-63.822} t)} \Big/ 4\pi}{9.1 \times 10^{-16}}$$

$$= 1.098 \times 10^{15} \times \sqrt[3]{3 \times (10^{-44.5} - 2.1 \times 10^{-63.822} t) \Big/ 4\pi}$$

At $t = 4.76 \times 10^{17}$ seconds (i.e. now), $r_{now} = 0.98365$. Thus, $r^2_{now} = 0.9676$, where $r^2_0 = 1$.

---

[30] There is an adjustment of the exponent from -63.5 to -63.822, to exhaust the volume of the atom in exactly 10^{19} seconds.

The relative surface area, in general, is

$$r^2(t) / r^2{}_0 = r^2(t) = 1.2056 \times 10^{30} \times \left[\frac{3 \times (10^{-44.5} - 2.1 \times 10^{-63.822}t)}{4\pi} \right]^{2/3}.$$

The net flux outward, $F_{net} = 10^{23}(1 - R)$, where R is the relative surface area (producing graviton re-capture), is therefore

$$10^{23}\left(1 - \left(1.2056 \times 10^{30} \times \sqrt[3]{\left[\frac{3 \times (10^{-44.5} - 2.1 \times 10^{-63.822}t)}{4\pi}\right]^2}\right)\right).$$

Moreover, relative $F_{now} = 1 - 0.9676 \approx 0.0324$.

The next task is to find the absolute inward flux as a function of the age of the universe, and for this it becomes necessary to track the matter density function with time since 10^{17} seconds.

The radius of the universe at the beginning of the quasar era was about 20% of what it is now, at 16 billion years. Thus, using $R = ct$, one finds a radius of 3×10^{25} metres at that time. The radius of the universe at any time t after that era will be given as $(3 \times 10^{25} + ct)$ metres – assuming c to be quite constant since then - and the average separation between its remaining 10^{81} particles is

$$s(t) = (3 \times 10^{25} + ct)/\sqrt[3]{10^{81}} \approx 0.03 + 3 \times 10^{-19}t \text{ metres}.$$

While the quasar era was in progress, the clock rates fell to around $t = 0$, creating the critical flux condition that triggered the explosions seen as quasars. Starting at that era, it is an implied condition that net flux $F \approx 0$, and $F_{out} = F_{in} = 10^{23}$ gravitons at this time. In the life of the universe after the quasar era, its volume will increase by a factor of a million, so the average separation of particles will increase by a factor of a hundred after 10^{19} seconds. The per-unit-time-and-area incoming graviton density should drop as $1/R^2$ throughout the post-quasar age. After $t = 10^{19}$ seconds a critical situation should develop, where most atoms become exhausted of their gravitons, and disappear. The atoms remaining would see $F_{in} = 10^{17}$ and $F_{out} = 10^{23}$.

This all assumes, of course, that matter is spread evenly in space, like a gas. Upon reflection, however, this is far from the reality, as matter – in practice – tends to clump in galaxies. Matter has therefore stayed at approximately constant density within most galaxies since the Big Bang event. The galactic agglomerations, however, have separated by a factor of about a hundred in the lifetime of the universe, so producing the reduced average density. The key point, though, is that the atomic clocks of the emitting atoms in these galaxies will not have sensed a significant local matter density change in the life of the universe. Gravitational clumping will continue to stave this effect off, so it seems doubtful that a correction for this effect will prove necessary.

The only factor producing a clock-based red shift is therefore the time intervening between the emission and reception of the photon. This time lag just means that the emitting atom is seen at a moment in time when its own size was a little larger than now, and its graviton recapture rate slightly higher. That is, the atom is seen at a time when its clock was running a little slower than it would be now, and – presumably – than all local clocks also do now. It appears necessary, therefore, to set $F_{in} = 10^{23}$ in the equation $F_{net} = 10^{23} - F_{in}R$, and to simplify this equation to $F_{net}(t) = 10^{23}(1 - R)$.

The final time-dependent graviton flux condition that remains is

$$F_{net} = 10^{23}\left(1 - \left(1.2056 \times 10^{30} \times \sqrt[3]{\left[3 \times (10^{-44.5} - 2.1 \times 10^{-63.822}\,t)\middle/4\pi\right]^2}\right)\right).$$

for t measured in seconds since the Quasar Era.

Net graviton flux, of course, directly determines the atomic clock rate. The model also presumes that the emission rate does not follow a relaxation curve – i.e. that flux from an individual atom does not decline exponentially with time – but that basic quantum processes keep emission steady until the end. While this *might* turn out to be a false assumption near the end of an atom's life, the relative clock-rates will otherwise generally go as:

$$\frac{F_{net}(t)}{F_{net}(now)} = \frac{C(t)}{C(now)} = \frac{1}{0.0324}(1 - (1.2056 \times 10^{30} \times \sqrt[3]{\left[\frac{3 \times (10^{-44.5} - 2.1 \times 10^{-63.822}t)}{4\pi}\right]^2})).$$

Since a photon grows in wavelength in direct proportion to the *apparent* downward frequency shift, and since this frequency shift is a reciprocal effect due to a faster clock rate, the increase in wavelength simply measures F_{net} now as compared to F_{net} when the photon was produced.

Using, $\lambda_{reference} / \lambda_{emission} = C(t_{now})/C(t_{emission})$, the relevant result for $t = 4.76 \times 10^{17}$ (i.e. now[31]) is

$$\frac{\lambda_{reference}}{\lambda_{emission}} = \frac{0.03238}{(1 - (1.2056 \times 10^{30} \times \sqrt[3]{\left[\frac{3 \times (10^{-44.5} - 2.1 \times 10^{-63.822}t)}{4\pi}\right]^2}))}.$$

In this description, clock-based red shift, z_C, needs to be defined as

$$\frac{\lambda_{reference} - \lambda_{emission}}{\lambda_{emission}} = \frac{0.03238}{(1 - (1.2056 \times 10^{30} \times \sqrt[3]{\left[\frac{3 \times (10^{-44.5} - 2.1 \times 10^{-63.822}t)}{4\pi}\right]^2}))} - 1.$$

Red shift measures *relative* wavelength increase – which itself measures the *relative* clock rate increase - since emission. The red shifts calculated on this model are clock-based red shifts – *not* cosmological or Doppler shifts. That is, this red shift occurs without any peculiar motion of the emitting galaxies and without any expansion of the metric. Any additional red shift seen in actual measurements will be attributable to these other effects.

The standard Doppler red shifts associated with look back distances appear below in column 5 of Table 1, for comparison with the current model, in column 4. Look back distances are calculated as looking

---

[31] There was an earlier adjustment of an exponent from -63.5 to -63.822 (to exhaust the volume of the atom in exactly 10^{19} seconds), and there is now a small adjustment from 0.0324 to 0.03238 to bring the cosmological and clock-based red shifts both to zero at exactly 4.76×10^{17} seconds.

back from a current age of 4.76×10^{17} seconds. Between 12 and 14 billion years of look-back, the quasar era should appear, and the values of z should grow rapidly in both models near the *beginning* of that era, somewhere near 15 billion years ago.

The Standard Model has ascertained that H – the Hubble recession speed – is 22 (±5%) km/s of recession speed per million light-years of look back, to at least $z = 0.5$. This figure has been used in Table 1 calculations. The last three columns are included only to make a point. They deal with the red shift differences that would appear between the clock-based red shift and Doppler red shifts, if observed galaxies have red shifts that really do fall neatly along the Doppler curve. This model will subtract the clock-based red shift from the "apparent" Doppler red shifts, leaving the residuals as the "true" Doppler contribution. The 2nd-to-last column then clock-corrects this residual, with the last column converting the Doppler shift into an associated recession velocity.

If the observed red shift profile faithfully follows the Doppler profile, it would mean – in this model – an implied residual z for $z < 0.3c$, that seems to follow the power rule $H = 48\,T^{0.6}$, where T is look back time in millions of light-years, and H is Hubble's recession constant in units of (km/s)/Mly.

Look-back time (millions of years)	Age in the universe (seconds)	Recession velocity (Standard Model) ($v = zc$)	Red shift, z_C (Clock difference model)	Red shift, z_D (Standard Doppler Model)	Red Shift difference (assumed to be Doppler red shift), D_{now}	Clock-adjusted Doppler redshifts, $D_{emission}$	Associated recession velocity (in km/s, at time of emission)
649.35	4.56×10^{17}	$0.05c$	0.043	0.051	0.008	0.008	2290
1298.7	4.35×10^{17}	$0.1c$	0.092	0.106	0.014	0.013	3820
2597.4	3.94×10^{17}	$0.2c$	0.204	0.225	0.021	0.017	5180
6493.51	2.71×10^{17}	$0.5c$	0.741	0.732	-0.009	-0.005	-1560
9740.26	1.69×10^{17}	$0.75c$	1.764	1.646	-0.118	-.0427	-13600
11688.31	1.07×10^{17}	$0.9c$	3.262	3.359	0.097	0.023	6500
12337.66	8.67×10^{16}	$0.95c$	4.2	5.245	1.045	0.201	35400
12987.01	6.63×10^{16}	$1c$	5.666				
CMWB?							
16233.77	-3.62×10^{16}	$1.25c$					

Table 1: Comparison of Red Shifts on Standard (Doppler Recession) and Clock-Based Models

On the model proposed here, a Doppler shift could still occur in the reference frame of the light emitter, where the clock ran slower. A Doppler-shifted photon suffers all the same scale length changes as any other photon emitted from that source, so any function tracing the

Doppler contribution as a function of distance needs to be adjusted for this. Recession velocities at each look back distance can be calculated from the Doppler red shift formula,

$$z + 1 = \sqrt{\frac{1 + \frac{v}{c}}{1 - \frac{v}{c}}}$$

Thus,

$$\lambda_{emission} \sqrt{\frac{1 + \frac{v}{c}}{1 - \frac{v}{c}}} \longrightarrow \lambda^D_{emission} = \text{Doppler-shifted emission wavelength.}$$

The clock differences then further red shift the Doppler lengthened photon, $\lambda^D_{emission}$, which in turn produces the locally observed Doppler contribution, $\lambda^D_{now} = \lambda_{total, now} - \lambda_{clock, now}$. In short,

$$\lambda^D_{now} = \lambda^D_{emission} \frac{C(t_{now})}{C(t_{emission})} .$$

The clock difference simply expands the wavelength - and the increment to the wavelength - by the same scale factor. To obtain the correct Doppler contribution at emission, it is necessary to take a table entry in column 6 and multiply it by

$$\frac{C(_{now})}{C(t_{emission})} = \frac{0.03238}{(1 - (1.2056 \times 10^{30} \times 3\sqrt{\left[\frac{3 \times (10^{-44.5} - 2.1 \times 10^{-63.822} t)}{4\pi}\right]^2}))},$$

to get the entry in column 7.

Theorists of the late 1800's argued for a photon that stretches as the metric (measuring frame) of space increases in size. This is presumed to track a fall in vacuum density. A photon bouncing between two mirrors for the age of the universe should eventually red shift if the metric - within which the photon and mirrors exist – expands, detectable if the measuring device (normally a spectrometer) does not also expand by the same factor. The red shift therefore traces not so much the distance, as the time, that the photon has remained unabsorbed.

What remains problematic in the Standard Model is the view that only photons stretch in the growing metric, while all (other) sub-atomic particles apparently do not. That is, the wavelength shift is seen *relative* to the "fixed" size of the atom. Yet the photon – in the current model – is not a distended object at all. Indeed – it is no bigger than a quark. So, how can it stretch? Rather, this model contends, the atom shrinks, as do all rulers, and the "wavelength" inferred by the frequency of the photon grows in inverse proportion. Perhaps the universe is expanding, but the observer is also shrinking! The effects are corresponding.

The red shifts calculated in this model are a different kind of "cosmological" red shift, predicted because of changes in clock rates, and this basically tracks changes in the surface areas of atoms. This does not yet include any effect for an expanding metric, other than to speak of a shrinking ruler as somehow equivalent to an expanding metric. The atoms shrink slowly in this model because they radiate gravitons. The loss of gravitons does not constitute a loss of mass because gravitons carry mass away, but rather because their emission produces a small decrease in the surface area of the atom. The continued emission of volume-occupying gravitons should, by the same token, act to drive the volume of the universe up, and – at minimum – to counteract gravitational collapse. The model assumes that this effect drives the overall peripheral expansion of the universe at speed c, although the expansion rate may seem smaller on shorter scales.

A universe's fixed differential growth rate dR – when expressed as a ratio of present radius R - is always therefore a decreasing quantity. The rate of slowdown is precisely what is measured by the universal gravitational constant, G. Some Hubble expansion should still occur, but it will be much smaller than generally believed, and the local Hubble constant should also fall with time.

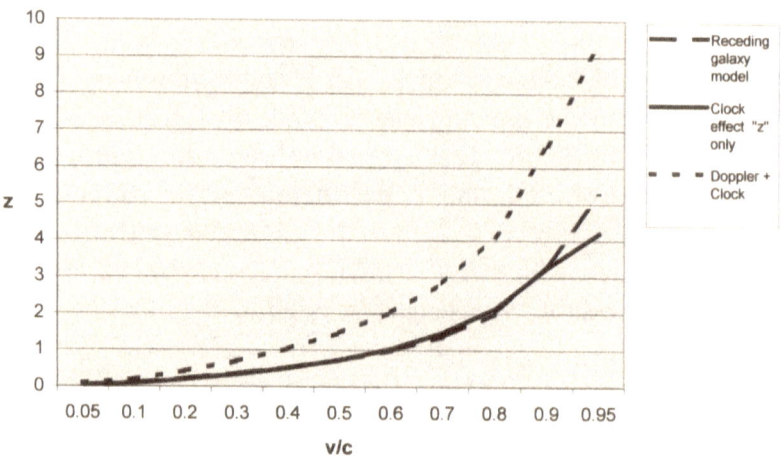

Chart 1: Clock and Doppler Red Shifts

Chart 1: Comparison of Red Shift Models

The curious fact in any model is that in the time t that it takes a photon to travel from source to receiver, the universe's periphery will recede a further distance ct. The photon travels, and the boundary of the universe recede, at the same speed. A photon emitted from the very "edge" of the local observer-centred universe at distance R would reach the local observer after time ct, but the universe would grow to size $2R$ in the same time, and its volume would increase eight-fold. The vacuum density – to follow the argument of the Standard Model – should, at final reception of the photon, have fallen to one eighth of what it was at the time the photon was emitted, suggesting a Standard Model red shift limit at ~8. This may in part explain why the universe starts to become empty of quasars and galaxies after z ~ 5. A few faint objects loom out from greater distances. The faintest quasars occur at around $z = 6.4$, and a few faint galaxies have been detected near $z = 7$, but these red shifts probably include large Doppler and gravitational effects, added to a cosmological shift nearer to $z = 4$.

---

[32] z ~ 4 sets a distance of around ~14.3 billion years look back, or ~4.365Gpc.

[33] That is, a measuring frame, or metric.

The clock-based red shift model does not expect to see any matter beyond the quasar period, so expects to see a cosmological red shift limit[32] near $z \sim 4$. But in the Standard Model, a block[33] of space can recede from the local observing frame at whatever speed it wants[34] – indeed it can recede at more than c. Therefore z - which would otherwise be limited to about 8 – could increase to infinity. The CMB radiation is formed – in some versions[35] of the Standard Model – near $z = 1100$.

It can be seen in Chart 1 that there is a divergence of the two graph lines at $v/c \sim 0.91$ on the Standard Model, as this is where relativistic effects begin to dominate in the Doppler formula. Red shifts will head off toward infinite values at the quasar era in the clock-based model, due to the critical graviton flux condition that triggered the quasar era.

Even Edwin Hubble – who first measured the distance/red shift relation – did not believe that the red shifts he saw were Doppler effects due to motion, even though this became part of standard dogma after him. He believed that a dimming effect should have been seen in galaxies as their red shifts increased. This effect was not seen, and if he had corrected the galaxy distances to account for the expected dimming, he would have largely lost his linear relationship between red shift and (corrected) surface brightness. The Hubble Law works well for *uncorrected* galaxy brightness. Hubble's lectures showed the Hubble Law relation for a universe *not* thought to be expanding. The clock-based red shifts in Chart 1 similarly refer to a non-expanding universe.

So far this model seems to say the following: all red shifts that conform to the standard Doppler profile associated with receding galaxies in an expanding universe *also* conform to the clock-difference profile for a graviton-radiating atom universe – but *without* any significant Doppler

---

[34] *Within* that receding frame, however, the restriction $v < c$ remains.

[35] The exact value of the red shift for the source of the CMB radiation depends on whether one allows H – the Hubble constant – to be an evolving figure or not, and this depends upon any time-variation assigned to the universal acceleration parameter.

contribution. If observed red shifts conform to a standard Doppler profile, this model can only interpret this to mean that the universe is little more than Steady State. Any recession velocity must be quite small.

The curve labelled "Doppler + Clock" describes what should be measured if both the clock effects and Doppler recession velocities exist in their predicted amounts. The two effects should each contribute about 50% of the observed red shift. If the observed red shifts – which seem at the moment to conform to the Doppler profile – are indeed really half clock-effect and half Doppler effect (these effects being approximately equal) – then it seems that the *actual* distances to the galaxies are almost exactly *half* of what they have been assumed to be. The universe is also about half as large as it has been assumed to be.

The Hubble constant – which quantifies the rate of increase of recession velocity with distance – does not change under these circumstances, though, as the recession velocity and distance both halve. So this partitioning of the red shift would not show up in the value of the Hubble constant. This splitting of red shift into half clock and half Doppler effect may also allow Hubble's surface brightness problem to be alleviated, as no more than half of the dimming he searched for would ever be observed. The apparent over-brightness of more distant galaxies compared to nearer ones would be explicable if they are actually only half as much further away from the nearer galaxies than was previously supposed.

THE ORIGIN OF GALAXIES.
QUASARS AND ARP OBJECTS

Figure 12: Abydos Temple Image
Found on the walls inside a 3000-year-old Egyptian temple at Abydos,
several hundred miles south of Cairo. If not a modern forgery, helicopters
and various other flying machines seem to be depicted. A History
Channel documentary explained these images as a coincidence, arising
when one dynasty partly chiselled away, and partly plastered over, the
carvings of another, adding new carvings, so that when the separating
plaster later fell away the merged carvings producing these images.

Quasars are extremely bright, but so distant that they have a star-like
appearance (hence, quasi-stellar object = *quasar*). But quasar red shifts
indicate huge look-back times (thus, huge distances) on local clocks,
and these objects are consequently considered to be about the oldest
visible objects in the universe. Their brightness can be a combination of
two things: they are intrinsically bright, and they are, on average, much

larger than galaxies, so have a large radiating surface. The latest images from the Space Telescope indicate that at the distance of the shining quasars there also seems to be a large population of faint objects. Look-back times correspond to the theoretical period in which galaxies were just forming. These faint associated objects could be proto-galaxies, but possibly also the remnants of other depleted quasars.

Observational evidence – under standard interpretations - is unanimous that the universe has gone into free-fall expansion. But the matter within it has the look of having been flung around by an internal explosion. The explosive force is usually attributed to the Big Bang, an event usually regarded as marking the beginning of the universe. The abundance of lighter chemical elements and the remnant cosmic microwave background radiation cannot be explained satisfactorily without the "hot fireball" description that appears in the Big Bang theory. But neither these, nor any other known facts, require the fireball to have been at the moment of birth of the universe. Any later time works just as well, provided it is no later than about 20% of the current age of the universe.

The matter in the early universe began at some point to collect into large proto-galactic clouds, which would later become quasars. Ten billion quasars may originally have existed. Within each quasar, up to 90% of the matter[36] could have been destroyed in explosions driven by matter dematerialization, with the remnant matter flung around in space to collide with other remnants to form galaxies. This explosive period was triggered by density considerations, and so probably occurred in most quasars at about the same time.

Probably the vast majority of today's galaxies developed from quasars. Each quasar had enough material to make many thousands of galaxies. Some quasars landed up, through later collisions, inside other galaxies. Some quasars remain linked by matter bridges to the latest galaxies they have spawned, and some have passed through galaxies and are moving

---

[36] Based upon the modern notion that only 10% of the necessary *closure* mass in the universe is seen, so the equivalent of the other 90% must reside in the form of energy or dark matter.

away from them, leaving gaseous trails. Still other quasars stand quite alone in space, nearly sole occupants of the voids from which they have violently expelled matter.

The quasar explosions actually helped stabilise the universe. Early matter clumps were uniform, unstructured masses, with virtually no rotation or turbulence. The quasar explosions were driven from regions near the centres of these gaseous clouds, where accumulated matter brought about critical radiation conditions. On the shock waves of each detonation, other detonations would have followed, with the fragments from these explosions being expelled in all directions. The collisions of these fragments with other fragments resulted in the galaxies.

Anything except head-on collisions would result in twisted agglomerations, with a mixture of linear and angular momentum. Nearly every galaxy observed has angular momentum, but it is not really known in current models how this momentum came to be. Astrophysicists produce thick books, containing dozens of possible dynamical models, in an attempt to model and explain observed galaxy structure, usually in terms of internal evolution and dynamical instabilities. But colliding high-velocity streams of gas are not generally discussed. Common explanations involve large gaseous clouds, with slight rotation, collapsing and flattening into much smaller discs with correspondingly higher rotation. But how the proto-clouds obtained their initial rotation is not explained. There is certainly no *one size fits all* solution in the Standard Model.

Understanding galaxies as the remnants of an explosive phase in the history of quasars *does*, by contrast, have the elegance of a one-size-fits-all solution. Rotating galaxies are dynamically more stable systems than extended, amorphous gas or dust clouds. The rotational inertia helps to locate the matter. It also slows the rate at which central accretion of matter occurs. Structures from the smallest to largest scales in astronomy show rotation, and it is evident that an enormous amount of localised angular momentum is distributed throughout the Universe. The universe must have a *net* angular momentum of zero, as the early universe had none, and this is a conserved quantity.

An interesting recent observation in relation to the detailed rotational dynamics of galaxies supports the scenario of dark clouds falling onto galactic discs. Most new stars – producing most of the blue luminosity in a typical spiral galaxy - are in the middle to outermost regions of the spiral arms, consistent with matter preferentially falling onto the outer parts of the rotating disc. If a galaxy is indeed surrounded by a diffuse cloud of ejected quasar material, such a cloud will be caused to swirl in rotation above the galaxy by the existing rotation and gravity of the galaxy. The result is that matter will be deflected outwards across the disc as it falls onto it.

Where matter falling downwards onto the galactic disc meets matter coming in from the edges, a corkscrew merger of the flows results, very conducive to producing the collapse of interstellar gas clouds and seeding star production. Thus, star birth is also explained much more easily by the continuing in-fall of mass at different angles than it is by random internal fluctuations within the rotating disk.

Another type of object that modern cosmology struggles with - but the current model deals with well – is an Arp object. An Arp object consists of a quasar and galaxy with quite different red shifts, yet connected by matter bridges. These objects – based on their respective spectra - *should* be extremely far apart, if the red shifts are cosmological. But if the excessive red shift of the quasar-like central object is due to a clock slowed by a critically dense matter core, observations conform to what is expected in the galaxy-formation model posited here[37].

If the quasar explosion does not fling matter away completely, a galaxy could form close to the quasar, giving a Seyfert galaxy. Such a galaxy may simply be a quasar seen through the "window" of a nearby foreground galaxy, giving that galaxy the appearance of having an abnormally high core flux. Or perhaps the dying quasar is finally entangled in another galaxy, and has a few final bursts of activity in which it dematerialises matter.

Any quasar can undergo successive periods of mass and energy ejection, if galaxies later collide or interact with the quasar, causing new density

---

[37]   If the high core flux indicates dematerialising matter, this means the clock has virtually stopped there, generating a huge red shift.

spikes. But most quasars are extremely distant, very bright objects, not visibly associated with any particular galaxy. Arp objects tend to be closer, and Seyfert galaxies are nearer again. This is all consistent with the hypothesis that the three types of object are progressively weaker sets of explosions during the lifetime of a quasar.

At some point in the future, no material will be left to fall onto the rotating galactic discs, and star formation will quietly cease. The local universe will become sedate and the night sky will dim. The universe will have a quiet end. Light will continue to arrive only from progressively greater distances, showing stars being born and radiating at earlier times.

The high-temperature processes taking place in the quasars produced every observed effect of the Big Bang, such as the original relative abundance of the light chemical elements, which certainly requires a very hot era in the early universe. The Big Bang birth idea cannot, however, explain how any kind of galaxy structure, or random velocity distributions, could have come to be, because a smooth linear expansion cannot generate the colossal distribution of angular momentum needed to gravitationally stabilise the structures seen in the universe. The standard Big Bang also has trouble accounting for the enormous preponderance of radiation over matter that is seen today.

To reiterate, the epoch of matter-creation is *not* signified by the quasar era. The ages inferred from cosmological red shift and velocity distributions of galaxies differ by more than three billion years. A whole range of approaches for estimating the current age of the universe lead to ages of between twelve and twenty billion years - a really quite *huge* uncertainty. The cold dark matter period, before the quasars, is *not* seen. The divergence speeds of galaxies, and the dimensions of the cosmic voids, allow the quasar period to be located approximately. This *looks* like the point of creation, but one should add up to four billion years to allow for clumping of matter to the critical radiation density required to ignite quasars. It should be clearly understood that galaxy velocities in the quasar period have nearly *nothing* to do with cosmological expansion. The galaxy velocity distribution generated in the chaotic quasar period is *superimposed* over a steady general expansion that was already in existence, which began a considerable time earlier and still continues.

The epoch in which the vacuum was compressed to form matter is of unknown length, but "switch on" probably occurred when the 10^{83} particles occupied a space with a radius of 10^{27} particles, or 2×10^{14} metres. This can be thought of as 30-50 solar systems of wall-to-wall atoms – about a hundredth of the distance to the nearest star. The actual switch-on time for the cold creation is ascertained by agreement of the cosmological red shift evidence for the most distant (quiescent) quasars and/or the oldest objects seen in any galaxy – Globular Clusters – taken with the radius of the universe at that time. Of these two approaches, that involving Globular Clusters is more reliable, as quasar fluxes may arise from glowing matter flung off at high velocities. Only stationary remnants of quasars are reliable, but the most distant quasars are so far away that they *all* look stationary. Globular Clusters are well understood, and the stellar models that predict what the evolution of a cluster should be like reproduce real measurements very well.

Based on the models, these clusters – objects formed even as galaxies were first being formed – go back as far as fifteen billion years, based on the 10% of these for which fairly certain results have been obtained among the 1500 or so clusters catalogued. Quasar activity probably extended right through from 95% to ~75% "look-back". Going back to when plasma instabilities set in, the universe reaches an age of 19-20 billion years. The Big Bang and galaxy creation phase date from around 15-16 billion years. The radius at the beginning of the quasar era was around 3.8×10^{25} metres.

Every observer will see the matter creation era as a thin shell formed at their own space horizon, and directly against that background the quasars, as just described in detail. The quasars are the most distant objects seen, because they were the first objects to appear after the Universe lost quantum mechanical degeneracy and became structured. Behind the quasars hide the dark, cold pre-quasar matter clouds, and then the quantum-mechanically degenerate era of matter creation. These distant, ancient clouds of matter cannot be seen as objects, but are there nonetheless.

LARGE SCALE STRUCTURE

45,000 BC China: Rock carvings of round UFO-like objects have been found in China's Hunan province. The depictions apparently date back to the age of the Neanderthals.

2345 BC China: In the *Hsui-nan-tzu* - a Chinese classic - there is a description of ten suns appearing in the sky.

9 BC Japan, Kyushu: Nine moons were seen in the night sky over the community.

Figure 13: Objects over Basle
Basle, Switzerland - August 7th 1566. The objects appeared at sunrise in the form of red, orange and black globes. After moving about the sky in an irregular manner, they faded away.

The era of smooth expansion probably ended when accumulating plasma had significantly filled the universe with electrified "lumps". That universe contained about ten times more matter than appears now. Those lumps coalesced into something like three thousand cells, with an average diameter of about half a million light years each. Each cell contained enough mass to construct hundreds of thousands of galaxies. Under continuous self-collapse, these cells eventually reached a critical density (about ten times the density of air), which then precipitated the quasar era – the true era of the Big Bang.

As each region reached this critical radiation density, enormous amounts of matter were suddenly converted back into mass, at between 50% and 90% efficiency. Huge amounts of ensuing radiation produced the great brightness of these objects. The explosion of one region triggered the explosion of the next, until all regions had - in a chain-reaction – seen the detonation of the matter located near their centres. Some of these centres of detonation are still seen as quasars, which became visible as the gas clouds were dispersed.

The intense heat associated with these explosions propelled large clouds of matter in all directions, generating the thermal spectrum now red shifted into the microwave background. Some of this matter still roams around as dark intergalactic clouds, and produces starburst activity whenever it collides with a galaxy. As these clouds encounter existing galaxies, they are drawn in spirally, landing along the bright spiral arms where young new stars are born. As little as ten percent of the primeval matter may have survived the quasar phase. More than half of this matter is possibly now located in the visible galaxies, and the rest in cold, dark intergalactic gas clouds.

Quasar explosions seem to account for the Swiss cheese (walls and voids) large-scale structure of the universe. Material ejected from simultaneously exploding quasars met in planes, producing "walls" of galaxies. Very few galaxies are located near the centres of the original detonations, which are now seen as "voids", but one would – for the more distant quasars – expect to see exploding quasars perhaps still in these voids, before they had dissipated themselves. The largest voids in space will identify where the first and largest quasars ignited. But in the case of short look-back times, some more recent quasars are still found positioned within the voids they have generated, although others have drifted out.

The universe is presently five times older than when the first quasar ignition occurred, and the average diameter of the nearby voids is about 7.5 million light years. This indicates that in more recent times massive explosions occurred about every 30 – 50 million years, pushing parts of walls back across voids. The most distant voids were probably larger, but they could also have been partially refilled by later quasar explosions. The periphery of a void - i.e. its total wall area – is now typically about 650 million square light years. A wall thickness of one to two million light years is also now measured, and the volume of the walls of each cell is typically observed to be about 1000 billion cubic light years. These walls hold the thousands of galaxies catalogued.

THE CMB RADIATION MAP

Figure 14: Strange Tassili Figure
Around Tassili, in North Africa, several hundred such drawings exist, scattered over many miles of desert: strange helmeted figures, some with antennae, some seeming to float as if weightless in a space walk. Other images depict what could be solar panels, space stations or floating spheres containing humanoid figures. Anthropologists explain the Tassili "roundheads" as ceremonial dancers wearing empty gourds over their heads.

The Cosmic Microwave (Background) Radiation (CMBR) seen today is an evolved record of an earlier epoch. At this earlier time, this radiation apparently peaked in the visible and ultraviolet parts of the spectrum. Expansion of the universe, according to standard theories, then shifted the peak of this radiation right over into the micro-wavelength band. Anisotropies in the spectrum are more pronounced at the longer wavelengths of the modern era.

After the explosive quasar era, any early record of background radiation *must* almost certainly have been over-written by the thermal record of the quasars. Wrinkles in background radiation observed by COBE – that is, non-random variations in the angular distribution of the radiation over the celestial sphere – should now map out the angular distribution of over a thousand quasar cells that filled the universe with their thermal signals.

The major problem finding of the COBE survey was the relative smoothness of the background radiation in *causally separated* parts of the sky. Some parts of the background sky should have been so distant from others that no communication – even by light-speed photons – could have brought enough density information across the universe to allow the inherent lumpiness to be smoothed out to the degree observed. The smoothness of the background is about an order of magnitude greater than would have been anticipated from random statistical clumping of matter. This is known as the Cosmic Microwave Background problem, and has prompted theoreticians to propose an early phase of ultra-smooth super-inflation in the universe, with a rapid subsequent slowdown to expansion speed c. But the quantum mechanism (the Higg's boson) needed to supply this super-inflation led to complicated additions to the equations describing the early universe.

It is simply *easier* to assume that the amount of matter now seen in the universe is an order of magnitude *less* than what it was at the time the microwave radiation was produced. With so much more matter at that time, the mass distribution was probably also an order of magnitude more uniform. Inflation theories may not be necessary to explain the apparent large-scale evenness of this radiation. Although quasars released most of their matter as energy, both the smoothness of the *initial* matter distribution, and of the thermal radiation produced as up to 90% of the mass of the universe dematerialised more or less at once, is probably now preserved in the smoothness of the CMB radiation.

THE EINSTEIN-PODOLSKY-ROSEN EFFECT

Figure 15: Object in the North Sea
This 17th century illustration depicts a sighting by two Dutch ships in the North Sea of an object moving slowly in the sky. It appeared to be made by two disks of different size. The source for this account is one of the books entitled *Theatrum Orbis Terrarum*, by Admiral Blaeu.

The Einstein-Podolsky-Rosen (EPR) paradox suggests the existence of information transfer at speeds well in excess of c. Particles leave a beam-splitter in complementary eigenstates, and are directed along different paths in space. Along one path a physical constraint is introduced, to bring about an alteration in the eigenstates of the particles in that beam. The remarkable observation - in an effect loosely called "quantum weirdness" - is that a simultaneous reciprocal change occurs in the eigenstates of the particles travelling along the other path. Logically, one might expect the two sets of particles to pursue completely independent futures once separated in the beam-splitter, but this weird correspondence effect continues to be observed in all forms of the experiment so far.

The effect remains a paradox if one *insists* that c is the maximum allowable speed for the transfer of vacuum information[38]. But in a model that does not suffer this restriction, this is no longer a paradox. Indeed, this experiment could as easily be interpreted as *verifying* for the first time that the vacuum has its own internal physics and velocity scales.

Special Relativity certainly forbids *energy* transfer in the vacuum at speeds in excess of c, even if parts of a wave could theoretically exceed this. The EPR paradox largely remains a paradox[39] because it is assumed a *boson* carries the information about the change of the conjugate state. The SR theory prescribes what happens within the *metric* in a given region, but really has nothing to say about *surface effects* in the vacuum, nor does it describe how vacuum particles sense each other's states. And yet – in reality – it needs these effects to allow the physical adjustments to be made in the vacuum that preserve measurement invariance.

The maximum possible physical speed for a *material* particle is c, because this is the expansion speed of the medium in which it finds itself. And this medium only expands in response to the introduction of new particles. Photons and gravitons, although non-material, have

---

[38] Quantum physicists would argue that it is a paradox only because physicists continue to insist that quantum physics follow Newtonian rules.

[39] Long debates have raged over this effect since the original paper in 1935 which argued philosophically, on the basis of (now-confirmed) predictions, that quantum states in one place could be known entirely from the measurements of their conjugates in another place. Experiments in which physical constraints placed upon one beam with predetermined states still resulted in "reciprocal"

no system of their own for propagating in the medium, so are also only able to move by virtue of merging with and being carried in the general expansion. This remains so without reference to any *intrinsic* properties of the vacuum medium.

Chief among the associated problems is that of establishing simultaneity of measurements, and therefore of quantifying what is meant by *instantaneous* conjugate response. "Instantaneous" can mean no more than some interval smaller than the smallest period that can confidently be measured on a clock. This naturally carries over into the problem of precision synchronization of clocks. Indeed, quantum physics makes a central tenet of the ultimate inability to state that two events are truly *simultaneous* – at least when measuring quantum events. This amounts to saying that the universe has no way to verify that events are simultaneous. One is always constrained to operate within error margins. This obviously blurs the issue of determining whether changes in one beam are "simultaneous" with those in the EPR partner beam.

But the appearance of a virtually *instantaneous* reflex might also simply indicate the rapidity of this vacuum surface action compared to the speed of light. It also suggests that the vacuum is highly sensitive to internal imbalances. Although the vacuum has activity, this does not mean that this activity is statistical "noise". In a truly noisy vacuum, it is unlikely that light or gravity could propagate without dissipation. Noise is a macroscopic description of random thermal fluctuations, but the vacuum is not chaotic in this way. Chaos reflects the "degrees of freedom" available to vacuum particles, and these are rather limited.

The difference between the speed of light and the vacuum surface speed can be compared to the difference between the motion of a crowd (photons in the vacuum) and the motion of an object passed hand-

---

conjugate states in the opposite beams, suggested a mysterious "backward" transfer of information at speeds greater than *c*. Einstein *et al.* used this idea– now called "entanglement" – to argue that quantum mechanics would need to follow some still unknown rule that harmonized the whole theory, and that quantum theory as it stood was still incomplete. Debate still continues as to whether the paradox is real or not. In any case, information transfer at speeds greater than *c* is an obvious way to resolve the paradox.

to-hand across the same crowd (vacuum "surface" speed). And so the EPR paradox may not be a paradox at all. Super-luminary speeds will not be easy to measure, though, and a clear theoretical approach will be important.

Any estimate of the vacuum surface speed must arise as a combination of fundamental constants, in the same way as the Planck or gravitational radii are calculated. To obtain correct dimensions, it seems evident that any characteristic quantum velocity, u, will necessarily be on the order of

$c^n\sqrt{}$(Planck length/Gravitational radius), which for n = 2 gives

$$u = c[h/mc]^{0.5}/[mG/c^2]^{0.5}, \text{ or about } 9 \times 10^{27} \text{ m/s.}$$

For n = 4 or n = 6 the speeds are 10^{18} m/s or 10^{13} m/s, respectively. This indeed suggests the possibility of very high vacuum surface speeds.

This result, as it stands, comes as a useful surprise. A speed of 10^{13} m/s is fast enough to traverse the current universe (radius 9.5 $\times 10^{25}$ m) in 3 x 10^{12} seconds, or 10^5 years. One would certainly expect to see quantum smoothing of the CMB radiation at all times, with this small crossing time. Thus, if this *is* the correct quantum-scale transmission speed, every event in the universe really *is* still causally connected with every other – an observation that inflation theories try to account for erroneously. This implies that the universe, at this stage, still has a strong geometric integrity, a result that draws the geometric and quantum descriptions several orders of magnitude closer together.

In terms of quantum physics, one cannot predict the future from even a *complete* knowledge of the present, and even obtaining this complete knowledge is impossible. The uncertainty principle described by Heisenberg, which underpins quantum statistics, embodies the empirical belief that it is not possible to completely know all of the parameters of even a simple system, because each new measurement interferes with and alters the system, rendering any previous knowledge useless. Clearly, what is essentially a *physical* problem can only be overcome if a means of *remote* (that is, *inferential*) measurement can be found. Faster-than-light vacuum speeds could, in principle, facilitate remote determination of states.

The Pauli Exclusion Principle, which describes statistically-based "selection" rules for fermions, serves to impose *order* on sub-atomic particles. Gravitons and quarks probably obey similar rules, and it is this kind of internal organisation that stabilises the universe. If the quantum universe was either under or over determined, the universe would be far less stable than it is. As it is, Einstein's view that "God does not play dice" still seems more true than its opposite.

THE TWIN PARADOX

Figure 16: Sego Canyon, Utah
Age estimated at ~7500 years. The image seems to contain tall, armless, dog-headed humans (or humans wearing hoods), and what look like lightning strikes and at least two flying saucers with ion-trails.

In the Twin Paradox, one of two young twin brothers is sent on a mission into space, from which he will eventually return. He is equipped with a wonderful rocket, providing him with relativistic speed, so relativistic effects will become important during his journey. After perhaps ten years he returns, and again meets his brother, but finds that his brother has become an old man.

This story is told so that people can understand, in simple terms, the repercussions of the Lorentz transforms for physical changes at relativistic speeds. The effect described is *not* an illusion due to motion, but a real effect that has been seen in laboratory experiments using particles with known half-lives. Effects associated with relativistic motion include the observer, at rest in space, seeing the travelling rocket become foreshortened in the direction of its motion, time on the same rocket slowing as its speed increases, and an increase in the mass of the rocket. Every effect has been confirmed in experiments with particles.

While the Twin Paradox provides a very valuable tool that can be referred to again and again in making arguments, it is important to be clear about when a situation really *is* the equivalent of that described in the paradox. The basic idea is that a measuring frame considering itself to be at rest really does measure these differences in the moving frame, even though no such changes are ever noticed by those within the moving frame. These are real effects that characterise the theory of Special Relativity, and are based on *slow* acceleration to relativistic speed – avoiding the need to correct for acceleration or gravity. Accelerated particles, and muons arriving from the external galaxy, are among examples that fit the description of *gradual* acceleration well enough for SR to apply accurately.

The Twin Paradox must be successfully married with the Principle of Equivalence – which appears in the theory of *General* Relativity – if there is a hope of applying the Twin Paradox to cosmological phenomena. SR insists that all reference frames agree about quantities measured in some other moving frame, irrespective of their own states of constant motion. To achieve this agreement, the Lorentz transforms allow conversion from one measuring frame to another, where *measuring frame* means a set of clocks, rulers and a means to measure mass. Therefore, *any* frame measuring the mass increase, time slow-down and length contraction in the moving frame must agree – on the basis of its *own* measuring instruments – on what it sees happening in the fast-moving frame. *All* observers agree that the clock slows, the rocket foreshortens and the mass increases, and they all agree on the degree to which these effects occur. This amounts to saying that the change in these quantities is no *illusion* due to motion, but is a real, agreed-upon, effect.

There is a certain one-sided aspect in this paradox. It cannot be resolved by saying that the same measuring effects are also seen by the travelling twin, so that as far as he is concerned, the clocks similarly run slow on Earth, the Earth is flattened in the line of sight and the Earth becomes more massive. Indeed, if all of these things are said about the Earth, it must equally apply to the whole universe seen by the traveller. Equivalence applied in this way really *would* produce a paradox, as the travelling twin sees the twin who stayed on earth age more slowly than he, and at their reunion each would have to be younger than the other. That is clearly impossible, and is also not in keeping with the experimental results.

In fact, the travelling brother witnesses the *opposite* of what his twin sees, if the effect really is *physical*. The traveller will look from the window of his rocket and see the mass of the Earth reduced in proportion to his velocity, so that clocks on the Earth are able to speed up. The Earth will also appear to elongate slightly in the direction of its apparent recession. The clocks on the craft do not seem to change, however, nor the physical dimensions of the craft. The Principle of Equivalence is more concerned, in fact, with what remains true in the observing frame than with apparent changes occurring beyond that.

There is a subtle physical reason for this difference in perspectives, which reconciles the two world-views as one. The reason is that a *closed* reference frame is being considered, and the energy being used to move the rocket at these relativistic speeds through space has been supplied from the energy available in the vicinity of the observer. From where else could all of this energy associated with relativistic motion have been obtained? While energy density vastly increases near the rocket, it is correspondingly reduced near the observer, in accordance with the principle of Conservation of Energy[40]. The energy in the system has simply been repartitioned. The local over-density near the rocket causes the clocks to run slow there, while the corresponding under-density near the observer causes those clocks to run similarly faster. Both observers see *real* effects.

The situation where two relativistic travellers look at each other has a few added complications. To reduce complication, they are taken as travelling in the same direction and at the same constant relativistic speed. These two simplifications of situation do not alter the physics, but they simplify the discussion for present purposes.

Rocket A has observers who see the whole non-relativistic universe *around* rocket B with its clocks running fast, distances elongated and mass decreased. Basically, it is a surrounding universe in which the energy density has all the appearances of having fallen markedly. This decreased energy density in the surroundings is, however, compensated by the increased energy density found associated with the relativistic rocket. Even though the observer in rocket A is *himself* travelling at relativistic speed, he sees the clocks in the other rocket slow down, and the mass rise, in the same way as a stationary observer would. The Lorentz transforms preserve this *true* result, even under these conditions. The transforms make no concessions to the fact that the observing frame also has relativistic speed – only requiring that its velocity be constant.

---

40   Importantly, energy conservation seems to run into a conundrum. To illustrate this, consider the case of a 100W, square, radiating surface on the earth, being observed by the travelling twin. Take a simple situation, such as that in which the rocket is viewing the Earth along the plane perpendicular to its motion, and let the rocket be travelling at that speed that sees its clock run at 75% of the clock rate on Earth. In this orientation, all lengths seen on Earth will appear unchanged. Thus the traveller might be thought to see the 100W unchanged. However, he sees the clocks on Earth running at 4/3 times their earlier rate, while the total number of photons radiated per second in the Earth frame remains a fixed number. The traveller measures a power rating of 133.3W, using *his* clock. If the rocket heads off instead in a radial direction from the Earth, all sizes on the Earth will grow by a factor of 4/3, and the radiating area will always seem 16/9 times larger than before. But the number of photons emitted into this new area is unchanged, so the power output is the same (although a large decrease in *intensity* is measured). Again, only the

Exactly the same description would conversely be made by the observers on board rocket B, looking across at rocket A, in accordance with the Principle of Equivalence. The fact that now *two* such rockets drain the surrounding space of energy to generate motion does not present any real problems. It simply means that both take longer to get to their relativistic speeds, explicable from the increased gravitational attraction each now feels towards the other because of their increased relativistic masses. On their mutually slowed clocks this change is not, in any case, noticed.

These two situations still remain *fundamentally* different to that in which *all* observers see the material just short of their own foreshortened space horizon receding cosmologically at just under the speed of light, with a clock that appears to run slow. Cosmological foreshortening is a *projection* effect, but should not be confused with the *real* foreshortening associated with relativistic motion through a *local* (i.e. constant energy-density) reference frame.

A basic difference that may help to differentiate the two situations is the fact that the mass at the horizon does *not* appear to increase to infinity. Secondly, the era being observed is one in which the laws of physics may have been operating with different values for the basic constants of nature than are now seen in the local universe. Thirdly, the Lorentz transforms applied to a local reference frame assume a constant energy density throughout the measuring experiment. The universe, however, has a different energy density at every time in its history, and this must also be accounted for when interpreting what one sees at the

---

clock difference is significant and the same result is returned as before. Thus, the Earth frame appears to radiate out more energy as the traveller's speed increases. The energy is *departing* from the earth frame to fuel the relativistic motion.

On Earth, an observer of a similar radiating square on the rocket in the first situation sees no change in its area, but sees the clock on the rocket running slowly. Thus the photons seem slow to emerge and a reciprocally reduced power rating is measured. The same reciprocity occurs in the radial situation.

While energy may not be conserved in either frame, the *geometric mean* of the energies in travelling and observing frame seems to be a conserved quantity.

local horizon. These all become complications for the observer now making and comparing measurements for phenomena at the horizon. Careful thought has to go into what horizon events actually mean.

As a first step in trying to understand the horizon situation and to clarify this difference, the evolution of the energy density of the universe will be considered, on the basis of the model for particle creation offered earlier. And, in making this argument, a clarification must be made about the way time and distance are measured in any locality, as a function of radial distance from an observer.

Because, ultimately, clocks and red shift are used to define distance, and there is no absolute clock, no absolute distance standard can be defined either (since $d = ct$, and t is arbitrarily measured on a local clock). This means that distance and time standards are purely relative to local energy density, and are always defined locally.

Now, if an initial cosmic compression originally pumped in all of the universe's available energy in the form of potential energy associated with matter, the history of the universe ever since has been one in which this matter slowly (or, at times, violently) releases its energy back into space. This process of dematerialisation is also driving the expansion - in effect, the relaxation - of the universe. If the universe, of radius R, expands at speed c, it follows that $dR/dT = c$, where it is presumed that c is essentially constant. The universe's rate of change of volume therefore follows the rule $dV/dT = (dV/dR).(dR/dT) = c(dV/dR) = c(d(4\pi r^3/3)/dR) = 4\pi cr^2$. Energy density consequently changes according to $d(\rho_E)/dT = d(E/V)/dT = [V(dE/dT) - E(dV/dT)]/V^2 = [V(dE/dT) - 4\pi Ecr^2]/V^2$.

If the total energy, E, in the universe remains constant (assumed and implied in the model), then $dE/dT = 0$, leading directly to the result that $d(\rho_E)/dT = - 4\pi Ecr^2/V^2 = - 4\pi Ecr^2/(4\pi r^3/3)^2 = - 4\pi Ecr^2/(16\pi^2 r^6/9) = -9Ec/4\pi r^4 = -k/r^4$ (where $k = 9Ec/4\pi$) at any age t after the expansion began. In short, this means that ρ_E falls as $1/r^4$ with time, indicating a time rate of change in reference frame energy density of $-1/r$, meaning that the overall energy density falls *hyperbolically* as the universe continues to expand. This is inferred by the fact that volume increases by r^3, while energy density falls by r^4. Integrating with respect to time, to get the time rate applicable at any given radius r, leads to the result that the local energy density falls and local clock rates increase according to $\ln(r)$ ($\pm$ an integration constant), out from the horizon. Since $r = r(t)$, then energy density also falls as $\ln(t)$, causing a logarithmic disappearance of "available" energy with time. This logarithmic drop in local energy density is what must be expected if the total energy is to be conserved in the universe with time, while it expands.

In going to the universe as a whole, there is a quite different effect, based on the Principle of Equivalence. Consider two observers located in different galaxies, in recession from each other due to the expansion of the universe, but able to keep sight of each other. Neither is moving locally at relativistic speed, so clocks are not seen to change their rates. There is no real *energy stress* in either observing frame. It therefore seems obvious that each observer must have a similar perception of the other. But what perception will each have?

The first observation must be that the *relative* rates of their respective clocks remain fixed, while the universe continues to expand, even if each clock runs at slightly different rates at the outset (due to localized energy density differences between the two galaxies). This holds true even if each clock continues to speed up with the course of time (due to the continuing expansion of the universe). Clocks will speed up slightly with time because the energy density (which essentially retards the clocks) continues to fall. The *ratio* of time-rate increase to energy density decrease will seem not to change.

No observer will measure this effect in his own frame. Nor can his neighbour in the other galaxy pick this change up, as his world is undergoing all the same alterations. So this amounts to another way in which the Principle of Equivalence applies, and continues to do so at all times, even if in reality the physical constants of the universe may be slowly evolving.

Eventually, the cosmic expansion will drive the mutual observers to the horizons of each other's universes. Locally, of course, nothing will ever appear to have changed. The universe will still appear to extend in all directions, clocks will still seem to run at the same rate as always and the total energy of the local system will always appear to have been conserved. Nevertheless, the energy densities *must* have decreased, because the universe has continued to grow, even if clocks appear to have ticked steadily. The photons emitted from each galaxy will appear increasingly red shifted with time in each receiving galaxy. This is *not* because the photons have lost energy, or because the vacuum itself has "stretched". It is because the clock rates in *both* galaxies have continued to creep upwards with time, while the light-delay continues to increase as the galaxies become ever more separated.

Of course, the energy density in the intervening space has indeed fallen, because the same energy is being spread into an ever-larger sphere as the universe rescales. So, while energy is *conserved* in the universe, the total *local* system energy could diminish with age in the universe, if the system being considered is not similarly rescaled. If not rescaling the local volume to retain a constant product for ρV, one would need instead to conserve the product of total energy and time rate (i.e. action) in each aeon.

As the observers in these hypothetical galaxies see each other approach opposite horizons, the distance between them seems to increase more slowly, if going by visual guides. A nice way to imagine this is to connect the two with a line representing a continuous stream of photons. While the galaxies are close to each other, the locus of the photon paths seems to be a fairly straight line across space, which one would regard as describing a flat (i.e. non-curved) space. But progress of time drives the galaxies apart steadily on their individual clocks and the geodesic becomes more curved as the separation increases, even if local clocks always seem unaltered.

Imagine a fictitious force now acting to slowly draw the two galaxies back together. Photons continue to be exchanged between the two galaxies, as before. The curved geodesic seems to straighten out again as they approach each other. The rest of the surrounding universe is allowed to expand all the while. At every stage the observers in the two galaxies have clocks running and changing similarly in rate. Each sees the clocks on the distant galaxies begin to "catch up" with their own as they draw closer. When they finally meet again, they agree that for the relative rates at which their two clocks were originally going when they first separated, the two clocks still agree on the relative ages now acquired by the two galaxies.

The time differences, when they were observed, had nothing to do with relativistic mass increase retarding time. They arose simply because at greater separations each galaxy saw the clocks of the other running at an earlier time, and perceived a slow-down with increased "look-back" (as clocks are always speeding up slightly with time). Even if the galaxies have been allowed to get so far apart that they had recession velocities approaching c, no *relative* change in the clocks could have occurred. There is therefore no equivalence between this situation and that of the rest frame observer and the relativistic traveller, and horizon effects and apparent clock changes should not be regarded as anything similar to the Twin Paradox.

THE UNIVERSAL GRAVITATIONAL CONSTANT AND THE AGE OF THE UNIVERSE

Figures 17 and 18: Wandjina Petroglyphs from Kimberley, Australia. Dated at approximately 3000BC.

The rate of matter decomposition ultimately guides any estimate for the eventual age of the universe. This is quite independent of gravitational effects and the like, so the Standard Model's assertion that it is the gravitational evolution of the universe that will determine its age and fate is false, because of *why* this is believed. In reality, the evolutionary fate of matter dictates the gravitational history and age of the universe.

One has to start with the question of how, in the process of time, matter will finally be returned to energy. Its total reversion to energy is assumed from a purely thermodynamic point of view. All systems seek to relax. It is difficult to imagine an energy-dense quantum state that will not evolve in time, whether it is an atom or the universe as a whole. The expansion of the universe is attributed to the fact that all atoms are being *unpacked* quantum mechanically of their gravitons, so that the universe as a whole is relaxing.

Secondly, throughout what has been written above, a distinction has been made between matter and the energy found in association with it. Rest mass is that numerical quantity used to give a *measure* of an amount of matter at rest, and thus of the amount of space which was compressed or configured to form the matter registering that mass. Inertial mass also includes what is added, by virtue of motion, to the rest mass. When matter is destroyed, this compressed space returns to the vacuum as rapidly as possible, as does any kinetic energy, and the effect on the vacuum is indistinguishable from that caused by a huge amount of gravity appearing in a short time[41].

In making this statement, there is a reassertion of another recurring theme so far, namely that gravity – while it draws objects together locally as a *side-effect* – is principally responsible for driving the expansion of the universe, and when it appears suddenly it is always explosive and volume-increasing.

Whenever *mass* is lost, it is in reality the spin angular momentum of all the vacuum particles found in association with and configured by the matter, of which that rest mass forms an indirect measure, that is scattered around into the vacuum. As matter is converted back into momentum-bearing volume elements in space – i.e. into photons - the mass appears to be liberated as *free energy*, which then becomes distributed broadly in terms of the quantum redistribution of spin states in the vacuum. While this may be equated with *energy* that is acting upon the vacuum, the vacuum actually responds to an infusion of free energy in the most efficient way it can, which is by local expansion.

Thirdly, while continuing to argue for the constancy of the speed of light as witnessed from every reference frame - because the expansion of the universe is ubiquitous - it may nevertheless be possible to allow

---

[41]   Hence the corresponding assertion in the Standard Model that *gravity waves* are emitted when matter falls into a Black Hole or onto a cosmic *string*.

the universal gravitational constant, G, to vary with time under some circumstances. It is assumed that – in the vacuum – the smallest constituent particles *cannot* be compressed, as this in principle attempts to divide space into volumes smaller than the quantum volume element occupied by the constituent particles. Therefore, h – which forms part of the definition of this quantum volume element - appears to be a constant, based on the nature of the vacuum.

By contrast, G quantifies the ratio of several variables at once, namely force, distance and mass. It seems possible that G could, in principle, be found to vary with the age of the universe, except in the case in which it is G that fixes the ratio in which these quantities co-vary.

It is another question, though, as to whether one could ever measure such changes, as these may follow in step with scale changes in the universe. That is, the problem could arise in which the *ruler* grows in exactly the same way as some fundamental length, or with any constant that expresses gravitational coupling forces in terms of lengths. But neither SR, nor GR nor QM present geometrical or physical reasons that insist this quantity should be invariant to preserve other physical laws. This gives some latitude to this question.

A good place to start this discussion would be with Newton's Law of Universal Gravitation itself. For a non-relativistic (low-energy) situation, two masses m_1 and m_2 feel binding force measured as $F_g = Gm_1m_2/d^2$, where d is the distance separating the centres of mass of the attracting systems. The current value of the "universal"[42] coupling constant, G, is 6.67×10^{-11} Nm^2/kg^2, or 6.67×10^{-11} m^3/kgs^2, in equivalent dimensional units. Mathematically, this comes out to be a very weak force, if compared to any of the other three basic forces currently known about.

The dimensional units can be reconstructed equivalently as $[m^3/kg][1/s^2]$ or $[1/\rho_M][1/s^2]$, which indicates that the gravitational coupling can in some way be interpreted as a relationship between *local* matter density (ρ_M) and a *local* differential of a gradient in time, because $1/s^2 = - d[1/s]/$

---

42   Quote marks are added, because experiment has not in itself shown that it is anything but a *local* value.

dt. These quantities appear to be the same *anywhere* in the universe at a given moment of local time, as all viewing frames are physically identical in the GR universe. Of course, when receiving information from objects at the horizon, one cannot be *sure* that current figures applied then, unless one has a firm physical grasp of why the gravitational force exists in the first place, and has thought through all repercussions of fundamental quantities – such as clock rates – changing with age of the universe.

Physicists do not usually think of the *local* universe existing in *any* time gradient, because one of the notions incorporated in the idea of locality is a *uniform clock*. But it cannot be denied that the expansion of the universe *must* be changing the universal matter density at all times. Thus – within the interpretation given in the current model for gravity - time must *universally* speed up by microscopic amounts at every moment, in response to ongoing changes in atoms and in matter density. The value of G is indeed one way to measure this rate of change.

This effect is necessarily the same in every reference frame. Nevertheless, this gradient is so slight as to appear flat for all intents and purposes, and this is why any Lorentz transformations still work to a high degree of accuracy (obviously up to 1 part in 10^{10}). The *constancy* of G will describe the constant change in speed of all local clocks relative to the age of the universe. And this in turn maps the rate of change of atomic radius. If the value of G does change *measurably* with age in the universe, it will need to be above and beyond the fact that the rate at which time flows in the universe is also changing with its age.

To identify possible relevant factors, Newton's equation will be checked. Force is linked, in the first instance, to the masses that interact. Mass is assumed to be a conserved quantity in every non-relativistic reference frame, a result following simply from the relation $E = mc^2$ and the fact that E is always conserved and c remains a universal constant. But energy is also described by $E = hf$, and changing clocks cause an apparent change in f. Thus rest mass energy E does in fact change with time. Therefore the product $m_1 m_2$ must also vary with the age of the universe, and lack of preference for one mass or the other implies that both m_1 and m_2 decrease in the same way with time. This correlates simply with the decreasing surface areas of the nucleons of the matter making up the interacting masses. A shrinking surface area is a shrinking capture cross-section, and this results in a declining gravitational interaction.

The force is also linked to the distance between the gravitating masses, as a $1/d^2$ term. The fact that it is square term simply relates how gravitational flux density falls on the surface of a sphere as the sphere grows to larger radii. *Unlike* other types of flux, however, the flux produces a *loss* of momentum in the line-of-sight between two interacting masses. As momentum represents a *stress* in space, causing motion, its subtraction must represent a removal of stress. Thus, when masses move toward each other under the influence of the gravitational force, they are really moving in such a way as to relieve a local stress.

The notion of distance, however, remains unaltered forever in a reference frame, as long as one moves at non-relativistic speeds[43]. So, like mass, distance does not seem to vary at all. Thus, the whole ratio $m_1 m_2/d^2$ appears to stay the same at all times and ages, for representative masses m_1 and m_2. This is quite irrespective of the fact that energy or mass densities may, as a whole, be reducing with the age of the universe. This ratio must remain insensitive to such changes.

If G changes with time, one can only apparently account for this on the basis of changes within the constituent atoms themselves. That is, if atomic dimensions do not change steadily with time, the value of G might be able to reflect *this*. Thus, if the cross-sections seen by gravitons trying to access the nuclei decrease with time, then the capture rate will decline, and the coupling strength will decrease. If one believes that mass is unable to vary, one will have to conclude instead that it is the value of G that is decreasing with time. Any corresponding increase in the d^2 term will have to be factored *out* of this, however, since this grows in tandem[44], and G will have to trace any residual change after this.

---

[43] For relativistic speeds, a relativistic correction is made anyway to Newton's formula.

[44] If atoms change their scales, then so will all rulers, so two fixed-position masses will seem to move apart merely because the *relatively* shorter ruler will think it sees an increase in the distance separating them.

For the sake of presenting a calculation, a model will need to be made. It will be assumed that gravitons are collected in something like a small orderly heap near the centre of a hadron, with the quarks in quantum orbits all about them. For ease of description, the action density associated with a graviton is regarded as residing *on* the quarks, and the total volume of all of the confined gravitons is seen as being *distributed* over the volume of the hadron as a whole. The atomic particle holds about 10^{42} gravitons, which will take between 10^{11} and 10^{12} years to sequentially escape their confinement. As each graviton is released, the volume of the system scales according[45] to $\Delta R/R = \sqrt{(\Delta M/M)}$, so that when the final graviton is liberated, the volume will fall finally to zero. The total initial volume occupied by this *pile* ($\sim 10^{-97}$ m^3) is about *forty* orders of magnitude smaller than the volume of the whole hadron, and so the physical cross-section of this pile will effectively be zero.

Imagine that *one* of the 10^{42} gravitons is released. Having occupied a volume of only $10^{-97}/10^{42} = 10^{-139}$ m^3, it expands on its exit by a factor altogether in the order of 10^{42}, making for an exit volume of around 10^{-97} m^3. This represents a particle with a radius of $10^{-32.5}$ metres - a few times the Planck length - that in the current model is defined dimensionally, without scaling constants, as $\sqrt{(hG/c^3)} \approx 10^{-33}$ m.

A relative volume decrease occurs in the hadron, on the order of around one part in 10^{42} for the original dimensions of the atom. The volume of the hadron at time 4×10^{17} seconds (i.e. now) is about 10^{-42} cubic meters. The volume is expected to fall linearly to zero in $\sim 10^{18.5}$ seconds. The cross-section of the atom is around 10^{-29} metres. Every second, for the $10^{18.5}$ seconds for which the universe is expected to exist, this is a *relative* coupling cross-section decrease of around $10^{-29} \times 10^{18.5} \approx 10^{-10.5}$ per second. The gravitational force, F$_g$, should be expected to decrease at about this rate with time, and the value of G = 6.67×10^{-11} tends to substantiate this prediction.

The rate at which time speeds up should *combine* the rate of decrease in local matter density with a slow decrease in the atomic capture area. According to the earlier model, at the end of the universe time will be running at close to its optimal rate. Not that this is perceptible. Time and

---

[45]   The rule R = $\sqrt{M}$ is established in the next two sections, and the differential form is used here.

lengths all change in tandem in any situation. However, it does stave off the problem of explaining why matter at that time will appear to become curiously less massive with time, as if dark matter in the form of exotic particles were suddenly being eliminated at the close of the universe.

Dimensional analysis provides another way to understand the meaning and significance of G. Using the fact that the units of G are Nm^2/kg^2, or m^3/kgs^2, and calculating in terms of volume per second squared rather than area per second, the calculations change in the following way. The relative *neutron relaxation* volume-decrease per emitted graviton $\approx 10^{-42}$. The total number of such emissions is, of course, $\sim 10^{42}$. The total relative volume change with emission of all gravitons is necessarily $\approx 10^{-42}$ x 10^{42} $\approx 10^0 = 1$. The total time needed to complete all emissions ($\sim 10^{11.5}$ years x 10^7 seconds per year $\approx 10^{18.5}$ seconds) is the estimated final age of the universe. The square of this time is 10^{37} s^2. The relative volume-change per atom, per square of elapsed local time, is accordingly $\approx 10^{-37}$ m^3/s^2. The total number of atoms in one kilogram is ~ 6 x $10^{26} \approx 10^{27}$. The total *relative* change in matter-occupation of space per kilogram of atoms is therefore $\approx 10^{-37}$ x 6 x $10^{26} \approx 6$ x 10^{-11} $m^3/kgs^2 \approx$ G. This result naturally agrees with what was just found.

However, this development carries a subtle barb. It indicates that the force of gravity – which is a quantity involving the scale factor G, would be *zero* if the cross-sectional area of the atom does *not* change with the age of the universe. To check whether this conclusion could be right, sense must be made of this mathematical assertion. That is, an explanation must be given for *why* the force would be zero if the cross-section did not change, within the context of this model. Luckily this can be done.

If the cross-section did *not* change, then the *net* change of graviton number in the nucleus must be zero. That is, as many gravitons would be entering, as leaving, the nucleus. Or, else, no gravitons leave or enter the nucleus. This is an equilibrium state that can *only* be set up in a universe where no matter is moving relative to any other matter, and every emitted graviton is recaptured by another piece of matter. If such a universe existed, obviously no objects would be able to feel any gravitational force, or else the static equilibrium could not stay static. All material objects – not coupled together by electrical or magnetic forces – would simply fall apart. Planets and stars would revert to large inert matter clouds, and stars would switch off. Thus, the force of gravity must be zero in that situation in which no relative motion exists. The solution has internal consistency.

In conclusion, it is really necessary to convert Newton's Law of Gravitation from the familiar form $F_g = Gm_1m_2/d^2$ into $F_g(T) = [dT/T][m_1m_2/d^2]$, where T is the current age of the universe[46]. Presently, and perhaps always, dT/T = $G = 6.67 \times 10^{-11}$. Since all *lengths* also alter at this rate (see *Why Does the Universe Expand at Speed c?*), it would hopefully be possible to determine dT (and therefore the universe's age, T) just by determining $d\lambda$ or df in a light ray. But what unfortunately makes $d\lambda$ and df appear to be *zero* in any interference or diffraction experiment is the fact that *all* lengths – including those of the grating - increase in tandem with the wavelength.

The problem remains of how to isolate a clock from the rest of the universe, and thus to halt its *own* ageing to allow differential measurements of time rates. Only by placing a clock close to a massive object could one slow its rate down and produce *some* of the effect of freezing it in its ageing process. Even then, it would *continue* to age, so any measurements would need to be done in terms of differentials. This remains a nice technical problem for someone designing such an experiment. Currently, a fixed-speed relativistic particle is the best approximation to such a slow clock. But even this particle continues to age.

What seems to emerge, however, is the suggestion of a new way of estimating distance to galaxies from their red shifts. Rather than using Hubble's Law, it should be possible to form an integral that uses the universal gravitational constant, G, to find the travel time (on a local clock) of the photon whose red shift has been noted. The shrinking of atoms should cause a slow decrease in m_1 and m_2, if G – which tracks atomic radius decrease – holds steady. Therefore, a gradual change in F_g should manifest itself, and the force of gravity should seem to decrease[47] as the universe becomes older, creeping downwards as the size of all atoms decreases. The universe will – in its last days – become approximately Steady State.

---

[46] Or, since $\Phi = dT/T$, $F_g(t) = \Phi m_1m_2/d^2$.

[47] This, incidentally, means that the ratio of gravitational to inertial force acting on a moving mass would also have to decrease as the universe ages.

THE FATE OF MATTER

May 15, 1879: In the Persian Gulf, in a sighting of about 35 minutes, witnesses aboard the ship *Vultur* reported two very large "wheels" – with an estimated diameter of 40 meters, and about 150 metres apart – spinning in the air, and then slowly coming to the surface of the sea.

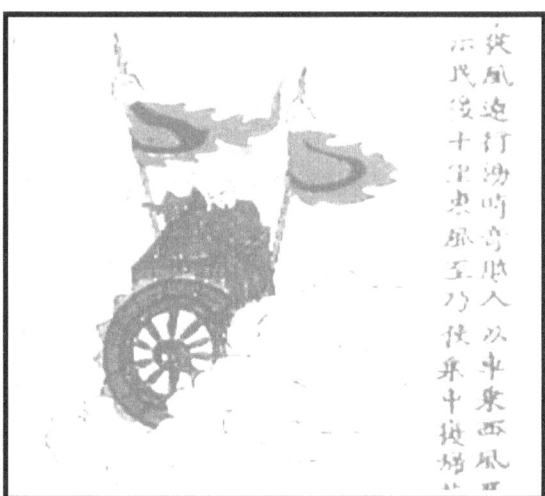

Figure 19: Flying Wheels - I
Illustration depicts the sighting of a burning wheel in the sky over Japan 900AD.

Figure 20: Flying Wheels - II
November 4, 1697: This picture records an extraordinary sighting over
Hamburg, Germany. The celestial objects were described as 'two glowing wheels'.

So far, the argument has proceeded on the basis of a hadron *fully* expelling all of its gravitons in its lifetime. However, nothing indicates that things are actually allowed to get this far. As the last graviton is expelled from a hadron, that hadron must vanish from space. But, long before this, instabilities might set in and cause the premature demise of the hadron. Indeed, as protons and neutrons expire, atoms will by stages transmute into simpler atoms, and finally into hydrogen, completely reversing the nucleo-synthetic process. As they transmute, they will release excess electrons, and the universe will become electron rich. As the hadrons expire, only free electrons and gravitons will remain.

If most atoms reach the end of their life at about the same time, a situation arises in which there is suddenly little left to drive the expansion of the universe. Gravitons are no longer being injected into the vacuum in an attempt to maintain its density as it expands. Cosmic expansion will steadily decline, and with the deceleration the speed of light will continue to fall, while the force of gravity between material objects will fade steadily to zero. In the end, material objects will seem to evaporate. Photons will gradually stop being emitted, and interact less often with matter. The universe will grow steadily darker. All physical processes will slow to a halt, and time itself will slow, until it finally ceases to exist. It should take about the same amount of time as the original length of the quasar era to see all of these physical and cosmological reversals take place, and to see all atoms fade from existence.

The current question, however, deals with whether this slow extinction could be pre-empted in any way by a catastrophic self-destruction of the atom at an earlier time. Just as the quasar era saw up to 90% of all matter being annihilated through a critical radiation situation, so it becomes necessary to check whether the atom could again encounter a critical radiation or matter density situation before its otherwise quiet eventual demise. The discussion will begin with the concept of critical matter density.

CAN BLACK HOLES EXIST?

The popular belief is that at a certain point - if too much matter is jammed into a given volume - that matter will undergo a catastrophic inward collapse, beyond the ability of any selection rule to resist, forming a "Black Hole". During this collapse, much of the material's nuclear energy is released, and the residue appears as the gravitational mass of a Black Hole. However, it is agreed that this also marks a point of departure from the *classical* Laws of Physics. Quantum physicists are not convinced that the universe can really allow this to happen[48], as there is the reasonable objection to some part of the original *vacuum* being destroyed with the dissembling matter. If it is not the vacuum that fills a Black Hole, then what does?

The theory supporting the existence of Black Holes attributes gravitational attraction to the presence of energy – whether in photon or material form. But in the current theory, attraction exists so long as atoms are able to absorb incident gravitons. Once the atoms are torn apart, this can no longer happen, and gravitons are instead liberated to drive an expansion of local space. Thus, a Black Hole should never

---

[48]   Indeed, in his paper *On a Stationary System with Spherical Symmetry Consisting of Many Gravitating Masses*, in the *Annals of Mathematics* (1939), Albert Einstein provided an argument to show that no such Black Hole singularity could be produced and remain consistent with the theory of relativity.

be able to form, and the Event Horizon should instead be an explosive surface. Also, *if* the vacuum still exists within the Black Hole, it too is presumably compressed to its maximum physically compressible limit. But this limit is presumably similar to that at which energy precipitates out of the vacuum as new matter, and this seems to present a contradiction.

To use a gravitational analogy from classical physics, an object always needs some *sufficient* velocity to escape from the gravitational pull of another object. This is the same, whether it is a rocket trying to leave a planet or a travelling object trying to escape enticement into some kind of orbit around another planet. In the case of a rocket trying to escape a planet, Newton's Law of Gravitation is equated to gravitational potential energy (mathematically defined with respect to an assumed vacuum potential energy of zero), to arrive at the simple relation $v_{esc} = (2GM/R)^{1/2}$. Physically, however, v_{esc} cannot be greater than c, and from this constraint flows the *critical* condition that $(2GM/R)^{1/2} = c$. This asserts physically that the associated radius within which some mass must be forced, to render it forever *bound*, is $R = 2GM/c^2$. The critical radius R is called the *Schwarzschild* radius. At this radius, a mass M is said to be at its critical mass density.

Near the limit at which this inescapable confinement is about to occur, clocks run wildly out of synchronisation, but all slow rapidly toward zero. Beyond this radius, matter and energy are swallowed up and *disappear* from the universe, as space is pinched irreversibly shut.

Since *mass* is a conserved quantity, the Standard Model argues that it must appear *equivalently* in the continuing gravitational presence of the Black Hole, so Black Holes produce extreme curvature in space, and forever hide the matter they contain from the universe outside. But the fact that the whole universe could in principle be gathered inside one Black Hole seems to present a logical anomaly. If this happened, the universe – which is defined by its mass – would end up having swallowed itself. The Black Hole would end up being "outside" the universe it contained, separating it from the universe itself.

THE PROBLEM OF NEGATIVE GRAVITON FLUX

The question that needs to be asked in the current model is what happens when matter congestion becomes so high that the *net* graviton flux for an atom seeks to become negative. Ordinarily, in this model, it is the net emission of one graviton per Planck unit of time per nucleon that maintains the cosmic expansion. When this stops, a miniature Steady State universe attempts to set itself up in the midst of a dynamic expanding universe that surrounds it. So, while most of the universe is still in expansion, a local constriction sets in, creating something like a quantum *back-pressure*, attempting to draw the vacuum back to fill the void developing around the graviton-depleted region.

The model allows re-absorption of gravitons, as well as their emission. From that point of view it might be mathematically possible for the graviton flux to adopt a negative value. The problem may not lie entirely with the atoms. These probably show a zero net change in volume as the gravitons are simultaneously emitted and absorbed. If the net flux continues to stay negative for many atoms at once, a significant slowdown of the local expansion will be felt. The velocity differential over the stressed region will quickly build. This all emulates a relativistic situation, where *inertial mass* begins to be added to rest mass, and the space around the moving object becomes tightly *curved* and distorted[49].

---

[49]   However, it does suggest that the relativistic model carries a small omission, namely that relativistic motion inhibits *the local expansion* from proceeding fully at c.

Physically - as has already been argued - time *stops* at the moment that the *net* flux drops to zero. In the case of just *two* gravitating masses, this can only occur if the two masses are interfaced atom-to-atom, which is just the situation in which they have merged into one mass. This makes sense, as *within* any mass the gravitational force falls to zero at its geometric centre. However, when *more* than two masses are in close association, the equilibrium-condition that *flux in = flux out* can in principle be met well *before* a full merger has occurred.

Each nucleon has a fixed rate at which it releases gravitons, and this is a limitation caused by the physics of the Planck pile. Something like one graviton can be expelled per unit of Planck time. To assimilate randomly arriving gravitons, the gravitons are probably admitted in an orderly fashion, and reassembled into the Planck pile. Ultimately, the issue goes back to the internal mechanics of a nucleon, and the hadron's apparent inability to absorb gravitons faster than it can release them. The inability to sustain a net negative graviton flux is tantamount to saying that the creation event cannot be recreated within the lifetime of this universe, or to saying that time cannot – for more than a few Planck moments – run in reverse. This may reflect a traffic problem: gravitons cannot travel both into and out of a hadron within the same quantum unit of time, or – at best – no more than one can enter while one leaves.

When more gravitons arrive at than leave an atom, the atom will tend to resist their assimilation. Most can only hover near the surface of the atom, or orbit the atom, in some sort of electrical confinement, but waiting to assimilate. For a while, they might queue in some sort of orderly pattern at the surface of the atom, the amassing total spin driving up the apparent mass. This emulates the relativistic situation. The *net* flux may *appear* mathematically to go negative (the same situation as when matter collapses inwards to form a Black Hole), but the reality must be that a shell of gravitons, akin to a shock wave, builds at the surface of the atom or hadron. The atom does not ordinarily expand or shrink under conditions of zero net flux emission, but as this shock region builds, the nuclear diameter will gradually become foreshortened. This is also a relativistic effect. The atom has *some* tolerance to such deformity, but it is not long before the atom undergoes an internal rupture.

In practice, the *critical* geometric condition arises when some piece of matter considered to be at the centre of some agglomeration sees *no gaps* in any direction through which its gravitons can finally escape. Because each surrounding piece of matter also radiates gravitons, each is in effect reflecting the gravitons received from the central mass. If *all* of the associated objects experience the same conditions, where gravitons continue to be produced within the agglomeration but cannot ultimately escape, the situation will soon demand a radical solution.

This is really a similar set of physical conditions to when material is travelling at speeds close to c. In other words, if one knows what happens at relativistic speeds, one also knows what happens near critical radiation density. From this equivalence, it is possible to assert that if one cannot travel at the speed of light, then neither should a Black Hole be able to form. The physical conditions met in the Standard Model are similar to those arising from critical radiation density in the current model, although these conditions occur at different matter densities.

CRITICAL MATTER DENSITY
IN THE STANDARD MODEL

In the Standard Model, matter undergoes a series of collapses to smaller containment radii, with each collapse relieving external stresses. Each such collapse also increases the structural integrity and lowers the total potential energy of the system, at the expense of loss of degrees of freedom.

First of all, there is electron degeneracy pressure, and when this fails to relieve growing pressure, the electrons are pushed back into the protons to turn them into neutrons. All particles are now neutrons, and pack further into a tight quasi-crystalline structure, gaining some stability from spin-pairing. The absence of electrical forces makes this a fragile state, and if the external pressure continues to mount, there is only a collapse possible into quarks, if these are indeed particles that can be set in arrays. But this is unlikely, and there is nothing known that would stop the final collapse to an object of zero volume – a singularity - as all remaining potential energy is converted into the gravitational mass of the Black Hole. In this theory, the *matter* disappears entirely, and a *hole* of radius cR appears in space where once the material object existed.

The Standard Model believes it finds confirmation of this step-by-step process in what appear to be electron-degenerate white dwarf stars and neutron stars, two stages on the path to the theoretical Black Hole. White dwarf stars result only from stars that originally held more matter than 1.4 solar masses (the Chandrasekhar Limit), and neutron stars *sometimes* result (fewer than expected have been catalogued so far) when a star exceeding five solar masses finally detonates. In both cases, enormous explosive changes of state are observed to occur first, leaving imploded stellar cores as super-dense *remnants* of stellar detonations (supernovae). It is *presumed* that the next - final - stage of collapse after the neutron star must be the Black Hole.

Clearly, the *larger* the amount of matter that collectively and simultaneously reaches the critical phase, the more extreme and catastrophic the methods that need to be employed to break the deadlock. The *speed* at which the critical mass is approached is another consideration, especially if it arises as the result of such a powerful and swift process as stellar implosion. Under these conditions, some time may pass before the system can react to the fact that it has gone *beyond* critical mass, yet the high vacuum surface speed may be the key to over-coming or avoiding this problem before it occurs.

When atoms shed their mass, they are merely returning energy – spin angular momentum - to the vacuum, in the form of photons. This is only possible if the matter is no longer present to contain these spin states. The material is destroyed or reconstituted in order to allow the release of this energy. The net result is that matter disintegrates or transmutes before it reaches the theoretical surface of the Black Hole, and energy is radiated as gravitons (gravity) and photons (light).

Beyond Critical Radiation

In the earlier discussion, the situation developed where the growing stress produced by a critical radiation condition resulted in relativistic surface effects on the atom. The question now remains of what ensues if that radiation pressure continues to grow without restraint.

The answer obviously relates to what happens within the nucleus of each atom, where all of the stress is being generated by virtue of the ongoing graviton production. One possible solution is that the atom simply *switches off* its graviton production. But if spirus pairs arise spontaneously, deep within the Planck pile, which is far beyond the range of the macroscopic world, external changes are unlikely to shut this process down. Quarks will continue to emerge from the Planck pile until a catastrophic situation becomes inevitable. It seems, therefore, that atoms – once starting on a net negative flux cycle – are soon headed for the equivalent of a nuclear *meltdown*.

The only two ways a system like this can reduce the building surface stresses are to undergo complete dematerialization, or else – as an interim measure - to try and relocate the material found in two atoms into one new atom with a new – but *relatively* smaller – surface area. After all, the largest surface area-to-volume ratio is associated with the smallest object, so smaller objects also feel the largest surface-stress-to-volume ratio. To reduce surface stress, it becomes advantageous to merge material with a neighbouring atom, to help offset imminent catastrophe. Thus, it appears - under conditions of extreme radiation pressure and space curvature - *nuclear fusion* sets in.

This is a nice solution. Several neighbouring atoms disintegrate, and their atomic particles reconfigure into a geometrically more advantageous arrangement. New, heavier, elements are born. The signal for this merger is carried on the surface of the vacuum, and the *coordinated* disintegration and reintegration of matter leads to brand new nuclei. There are now fewer atoms than before, because the fusion has increased the atomic masses through transmutation. Binding energy is released as nuclear energy, which is in the order of magnitude needed to release the atoms from their conundrum and to discharge the gravitons back into the vacuum in waves, removing the immediate stress. The newly formed heavy elements are also scattered by these explosions.

If the stress continues, this process will repeat itself over and over again, until the matter is all pushed apart on the shock waves, finally breaking the critical enclosure. This is nothing but the engine of cosmic nucleo-synthesis, and most of this took place in quasars, as the initial hydrogen clouds contracted gravitationally to critical densities.

In the Standard Model, Black Holes are intrinsically invisible, but can be identified by the matter that aggregates about them. Quantum mechanical considerations have also turned up the result that Black Holes – if they exist – must at least radiate heat. Observations of quasars and galactic centres confirm that large radiant energy emissions are found in association with the in-fall of matter, suggesting the presence of Black Holes. This photon flux is interpreted in the Standard Model as *synchrotron radiation*, which is light generated by the changing angular momentum of photons as they spiral into a Black Hole. This may account for much of the flux observed, but the Standard Model relies too heavily upon this interpretation. One should consider instead that the high fluxes seen from quasars or galactic centres could be due to stress-relieving bursts of nuclear fusion.

MODELLING CRITICAL
RADIATION

One way to model the radiation problem in the current theoretical approach is to think of an emitting body surrounded by partner emitters. If all emitters are spherical, for simplicity, with surfaces transmitting outwards all internally generated gravitons, then the condition of net-flux-equals-zero is achieved when the surrounding bodies collectively return the same flux to the surface as it emits.

The central body can be assigned a radius of one metre. Its surface area is 4π square metres, and its cross-section π square metres. The outward flux density can be appropriately scaled so as to equal one unit of flux per unit surface area, making the total outward flux 4π units in all. Now let fourteen similar spheres (with a *collective* cross-section of 14π square metres) be arranged so as to completely enclose the central sphere by surface contact in all three-dimensions. The central mass is therefore effectively enclosed by hexagons on a circle.

Let each emitting sphere weigh one kilogram. Let these be considered to collect *all* (in fact $\approx 3\pi/6\sqrt{3} = 0.9069$ is the relative capture ratio) of the flux from the central sphere. In this situation, the problem is that they radiate back far more than they receive, so need to be retracted to a distance at which the central sphere collectively receives from them the same as it emits. The change in distance causes gaps to appear between them, so the capture rate also drops with distance.

Each of these fourteen spheres must radiate back $4\pi/14$ units of flux in the direction of the central sphere. This occurs when they are set at a distance of $\sqrt{14}/2$ (= 1.88) metres from the central one. At this radius, the surface containing the fourteen spheres which radiate inwards has an area of 14π square units, and the 14 radiating spheres all present a face of area π towards the central sphere. Any six lay on a circle of 11.8 metre circumference, centres separated on average by 1.97 metres on the arc, meaning the masses are marginally over-lapped. The *geometric* arrangement is the key thing, keeping fluxes – at whatever common level these may be – in balance.

To see it another way, imagine the central sphere, in the first situation, emitting *one* graviton to each of the surrounding spheres at a time. Each of these surrounding spheres is surrounded in the same way, therefore similarly seeing fourteen spheres. Each of these is radiating in the same way. On average, each of the fourteen spheres emits one graviton in the direction of the central sphere in every fourteen emissions. The collective effect is that for every graviton emitted by the central sphere, one is returned, keeping the balance. No radiation loss is assumed. Thus, the *contact* situation – without further compression of the masses - is approximately the critical radiation condition.

In the real universe, the graviton fluxes assume specific values, accounted for in the coupling constant G and the actual masses present. Each of the fourteen spheres has potential energy (per unit mass) with respect to the central mass, calculated (through Newton's formula) as $gr = -GM/r + C$, where C is Newton's integration constant. In total, they represent per unit mass potential energy to the sum of $-14GM/r + 14C$. Any force exists to remove a stress, and in this case the force producing the negative potential difference exists in order to maintain the net gravitational energy of the system at zero. This assumes that an equal positive amount of gravitational energy is also somehow associated with the masses in the interaction.

To mathematically avoid allowing the energy to become zero in the external universe, it is necessary to assign the integration constant to be equal to GM/r. *Physically* this is equivalent to representing the *bound* energy – i.e. the rest mass – as c^2 per unit mass. In other words, 14C = 14GM/r = $14c^2$ in the relation above. In this way, one kind of critical density can be defined, which is that at which energy becomes matter, and vice versa. This is (within a factor of 2) the Schwarzschild radius for the Black Hole. Solving, it is found that GM/r = c^2, or r = GM/c^2.

In terms of the fourteen spheres, however, the separation they require to stave off flux imbalance is geometrically determined to be $\sqrt{14}/2$ units of radius. In other words, at a 14:1 mass ratio, the flux balance occurs for a $\sqrt{14}:2$ radius ratio, assuming that the fluxes are linked directly to their masses. Generalising to any mass M, an M:1 mass ratio sees the flux balance occur for a $\sqrt{M}:2$ radius ratio. So, for matter homogeneously radiating gravitons, and disallowing negative graviton flux, a critical radiation situation will develop when a unit mass of unit radius is surrounded by an agglomeration of mass M at mean radius R (in metres) of $\sqrt{M}/2$ (M, in kilograms). This is *not* a critical mass condition, but a critical *radiation* condition. This comes into effect *earlier* than the critical mass situation, and largely exists to help the universe *avoid* the physical catastrophe at the Schwarzschild radius.

To explain this with an example, consider a quasar carrying the equivalent mass of a thousand galaxies, namely 10^{43} kg (i.e. 10^{12} stars/galaxy x 10^{57} atoms/star x 10^1 atomic particles/atom x 10^{-27} kg/atom). Once this mass occupies a volume whose radius in metres is $\sqrt{10^{43}}/2 \approx 10^{21.5}$ metres - about the diameter of the Milky Way galaxy ($10^{5.5}$ light-years, at 10^{16} metres/light-year) – the critical *radiation* balance will be reached, and violent nuclear fusion will start up, trying to drive the material apart. Thus, when a thousand galaxies get pushed into the space of just one, critical radiation density is reached. Compared to this, the critical *matter* density occurs at R = GM/c^2 = 6.67 x 10^{-11} x 10^{43} / 9 x 10^{16} = 7.4 x 10^{15} metres, some five to six orders of magnitude smaller again.

The existence of the critical radiation density for gravitons, based on the relative inability of the graviton flux density to go negative, explains why the quasars suddenly ignited. In igniting, the quasars converted vast quantities of mass back into energy, created virtually all heavier elements, and propelled high velocity clumps of matter in all directions, with the merger of these colliding clumps creating the rotating galaxies. This is also where much of the *missing mass* is located. It went into the massive quasar light fluxes, into the large kinetic energies of molecular clouds, and into the rotational energies of galaxies. Through shock waves and repeated ignitions, a large fraction of the early matter was reduced to energy, and the Big Bang resulted.

CRITICAL RADIUS AND THE ATOM

2000 BC. Peru's pre-Incan civilization records that the gods were from the star system Pleiades. Their legends tell of spaceships that came from the stars. Incan texts reveal that the Incans knew the earth was round.

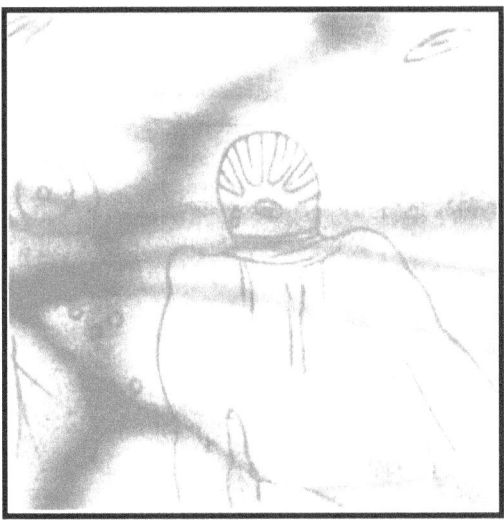

Figure 21: Tassili Spaceman
6000 BC: Image found in the Tassili Mountains, Sahara Desert, North Africa.

Nature has a means of staving off critical matter density: atoms dematerialise when the critical radiation condition is reached. This can be thought of as a safety valve, by which the universe ensures that conditions present *within* an atom are not replicated outside it. In short, it preserves the atom and the exterior space in separate worlds. Once these worlds merge, the universe is at an end, at least for that atom.

However, it seems that the situation deep inside the atom – at the Planck pile - is more like that of critical *matter* density. The pile of gravitons at its centre resembles a miniature Black Hole, radiating gravitons at a regular rate – just as the thermal Black Hole is believed to radiate thermal photons from its surface. This pile has dimensions about twenty orders of magnitude smaller than the Planck[50] length of 10^{-33} metres, and obeys a critical radius formula which has the same *form* as that seen in the exterior world, apart for employing different constants.

Using a dimensional argument, the characteristic quantum velocity u was earlier postulated to be $\sim 9 \times 10^{13}$ m/s, using the identity $u = {}^6\sqrt{(hc^3/m^2G)}$. But at less than the Planck length there is no significant interpretation (in my view) for the constant G, or for the ratio G/c^2. This ratio outside the atom is linked to the local potential energy of matter compared to its energy when it reaches the local horizon of the universe, and to the rate of change of cross-section of particles responsible for gravitational forces. None of this applies at the Planck length, or even generally at sub-Compton dimensions, where gravity is not experienced. The ratio G/c^2 presumably reduces to unity, as no time gradient exists within the atom (as it does in interstellar space) across which potential energy can be dissipated. The factor 1 presumes it is all dissipated already.

The critical radius condition has now simplified radically. That is, $R = \sqrt{(GM/c^2)}$ has become $R = \sqrt{M}$. The equation for critical matter density has simplified into the equation for critical radiation density. And this makes eminent sense, when considered. The quantum world near the Planck radius should therefore find a critical radiation

---

50   See Introduction for this work's definition of the Planck length.

balance with the world beyond the atom's "surface" when R = √M = $[1.67 \times 10^{-27}]^{1/2} \approx 4 \times 10^{-14}$ m, which is indeed close to the Compton length observed.

Within the Planck pile, however, it is less clear what the *masses* become when the interaction surface is so extremely diminished. A Planck mass of $(hc/G)^{1/2} = 5.5 \times 10^{-8}$ kg, associated with each of the 10^{42} spirii, would produce a total mass of 5.5×10^{34} kilograms, some sixty orders of magnitude greater than the 1.67×10^{-27} kilograms actually measured for the neutron. But this just illustrates the error of associating the idea of *mass* with the spirii. Mass measures the interaction of the entire hadron with the external world, and is linked to the energy that binds the quarks. *Quarks* are certainly "felt" in the dimensional world of the atom – but this is well above the physically invisible Planck radius.

Within the Compton radius, the value of G/c^2 is unity. This is actually required for R = √M to apply. Any particle that emerges from the interior of an atom must do so at the critical balance between the forces it feels and its size. Its *own* radius also obeys the relation R = √M. Emerging into the general expansion, it will adopt Planck length and mass proportions and accelerate to a velocity of *c*. To find the associated energy in Joules, and re-scaling so that the graviton is defined to have a "unit" of volume and contain a "unit" of energy, a scale factor k will be needed to re-scale the mc^2 term. Thus, $m_{pl} \approx (hc/G)^{1/2} = 5.5 \times 10^{-8}$ kg, and $E_{Pl} \approx km_{pl}c^2/2 = k(2.75 \times 10^{-8} \times 9 \times 10^{16}) = (24.75 \times 10^{8})k$ Joules, and this suggests that $k \approx 4.1 \times 10^{-10} \sim G$.

A dimensional analysis showed earlier that a speed of 9×10^{13} m/s might be possible in the free vacuum, and one should probably consider that such speeds are also seen within the Planck pile. If speed rises with spirii density, ρ, according to $u \propto c\sqrt{\rho}$, it suggests that the Planck pile density is as much as 1.5×10^{10} (i.e. 1/G) times greater than that of the quarks that make up the hadron. It is probably no accident that the ratio of the vacuum surface speed to the speed of light is about the same (about 10^{5}) as the volume density ratio for matter at the critical matter density to the critical radiation density.

A vacuum interaction-speed on the order of 10^{14} m/s, about 3 x 10^5 times faster than the speed of light, would effect a significant smoothing of the universe. This velocity represents a universe crossing time of 10^{12} seconds, or about 32000 years. Up until 10^5 years ($\sim 10^{13}$ seconds), a signal could have crossed the universe in less than one second, making the universe look rather like a single distended particle. Even at 10^{18} metres, the crossing time would have been just a few hours ($\sim 10^4$ seconds). This ability to convey density information right across the universe would further explain the smoothness seen in the CMB of the universe.

WHEN THE COMPTON RADIUS
REACHES THE PLANCK RADIUS

4000 BC: The Sumerians had contact with extraterrestrial civilizations, according to their creation text. The extraterrestrials interbred with humans, and the kings were taken to the stars by the extraterrestrials. The Sumerians say that the extraterrestrials came from Mars, from the star system Pleiades, and from the star Sirius. Sumerian texts show informed drawings of the solar system.

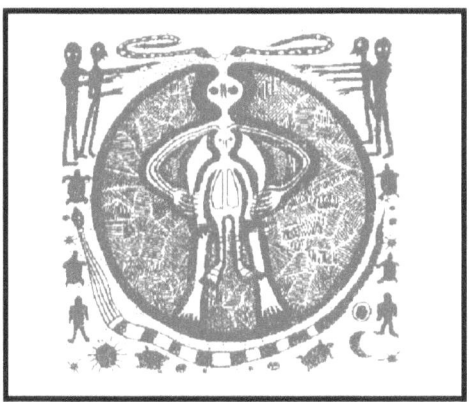

Figure 22: Extra-Terrestrial Cosmos?
In the image above, the figures running around the outside are probably astronomical or astrological emblems, while the larger central figures seem to depict a god and a human, in uncertain relationship. It may refer to seeding of the human race by extraterrestrials.

In deciding when and if the atom will become unstable, it seems that the best approach is to determine when the graviton-pile, deep within the nucleus, first becomes accessible to quantum effects seen in the macroscopic atom. The Compton radius marks the dimension at which quantization gives place to broad geometric rules, and represents the "surface" of the atom. The Planck pile has dimensions far below the geometric world, and remains invisible to it.

It seems sensible to assume that each of the 10^{42} emitted gravitons generates a similar decrease in atomic volume at its exit. If the Compton sphere of $\sim 10^{-45}$ m^3 is dissipated in 10^{19} seconds, about 10^{-64} m^3 of the Compton sphere must be dissipated each second[51]. One emerging graviton should therefore represent about $10^{-64}/10^{23}$ = 10^{-87} m^3 in the external atom, while its "effective" exit radius is approximately 10^{-29} metres. The Planck size is defined in this work as the geometric mean of the standard gravitational and Planck radii. Performing this calculation, the Planck size is set at $([mG/c^2][h/mc])^{\frac{1}{2}} = (10^{-53}.10^{-14})^{\frac{1}{2}} = (10^{-67})^{\frac{1}{2}} \approx 10^{-33.5}$ metres, so a Planck sphere occupies $\sim 10^{-100}$ m^3. The Compton atom becomes as small as the Planck sphere only in the last moments of the universe. One could therefore argue that the end of the universe is defined as that time at which the Compton sphere has finally diminished to the scale of the original Planck length.

But the death-knell of the atom could indeed come as early as when the Compton radius becomes smaller than the current orbit size of the quarks. This radius is conceivably located near the geometric mean of the Compton and modified Planck radii, which occurs near 10^{-24} metres. The volume (and thus number) of remaining gravitons at 10^{-24} metres is $1/10^{30}$ of that at 10^{-14} metres. So, even the geometric length of the quark orbits is not really reached before the end of the universe.

---

[51] Although this is the size of the Compton sphere *now*, and not at the birth of the atom, the current age (about 5×10^{17} seconds) of the universe is still just 5% of its projected eventual age, so the Compton sphere has diminished in size by probably no more than 5% so far. There are *still* $\sim 10^{19}$ seconds left in the age of the universe.

The gravitational constant, G, tracks the way in which the nuclear radius and atomic capture area decrease with time, in accordance with the release of gravitons and depletion of the central pool of 10^{42} particles. Each particle emitted represents a rest mass decrement of about 10^{-69} kg. Every Planck unit of time, a graviton is produced and emitted. This time unit is on the order of 10^{-23} seconds, so about 10^{23} gravitons are emitted every second. At this rate it will take about 10^{19} seconds to emit them all, which is between 10^{11} and 10^{12} years, more than twenty times the current age of the universe, if net emission remains approximately constant throughout its age[52].

With each emission, the gravitational and Planck radii shrink microscopically. The definition $dR_{grav}/dt = d[mG/c^2]/dt = G/c^2 [dm/dt]$ $= G/c^2(-10^{-69}$ kg/graviton x 10^{23} gravitons/second) leads to $dR_{grav}/dt =$ $-G/c^2 (10^{-46})$ m/s $= -10^{-63}$ m/s, using G = 1 and $c^2 \approx 10^{17}$. While $c^2 \approx 1$ in the Compton sphere, even further in – within the gravitational radius – it behaves more like material at the critical matter density, leading to the re-adoption of $c^2 \approx 10^{17}$. The internal pile consequently shrinks at around 10^{-46} m/s, which is about 10^{-13} Planck lengths per second. The gravitational constant, G, stays close to constant, because the rate of change of atomic cross-section – although a hyperbolic function - stays approximately linear for most of the age of the universe.

Since the Compton radius is set by $\sqrt{M}$, the atom's surface radius should undergo a (per second) relative radius change of $(\Delta R/R) =$ $\sqrt{(\Delta M/M)} = \sqrt{(10^{-69} \times 10^{23}/10^{-27})}$ kg/s $= \sqrt{10^{-19}} \approx 10^{-10}$ at every graviton emission, so if R = R_{atom} is about 10^{-15} metres, ΔR must be on the order of 10^{-25} m/s at the moment. Thus, the Compton radius shrinks faster than the gravitational radius in absolute terms (as $-10^{-25} > -10^{-63}$) at the moment.

A calculation shows that in the 10^{19} seconds of an atom's existence, the relationship $dR_{atom}/dt = -10^{-25}$ m/s would bring the Compton radius of the atom ($\sim 10^{-14.5}$ m) to zero after only 10^{10} seconds, which is just a 1000 years. It should, however, be noted that $\sqrt{(\Delta M/M)}$ implies a changing value for M with time, while ΔM is fixed at 10^{-69} kg per

---

52   This has already been shown to be the case, aside from current considerations.

emitted graviton. This equation conveys the property that mass must stop being emitted when M has declined to just 10^{-69} kg. But this is just the last graviton to be emitted, and just before its emission, R has reduced to ΔR, so R $\rightarrow$ 0 at the last emission. In the ratio $\Delta M/M$, the M keeps decrementing by ΔM, so an emission with ratio $\Delta M/M$ is always followed by an emission at ratio $\Delta M/(M-\Delta M)$. When just one graviton is left, $\Delta M = M-\Delta M$, or $M = 2\Delta M = 2 \times 10^{-69}$ kg.

M is the atom's original mass, from 5×10^{17} seconds ago. Since then, the atom has shed 10^{-69} kg x 10^{23} x 5×10^{17} = 5×10^{-29} kg – about 5% of its current mass of 10^{-27} kg. Thus the ratio initially equals $10^{-69}/(10^{-27} - 10^{-69})$ or 10^{-42}. This makes sense, as the first emitted graviton carries this proportion of the initial graviton pile at its emission. This ratio moves steadily toward a value of one at $T = 10^{19}$ seconds, at which point the process is ended. Initially $\Delta R/R$ must similarly equal 10^{-21}, and also move towards a value of unity after 10^{19} seconds. Both have the shape of a growth curve – but virtually flat for the age of the universe, and climbing steeply only at the end.

This is useful, as it allows a calculation to be made of where the linear rule for the gravitational radius meets the quasi-linear rule for Compton radius. The Compton and gravitational radius expressions, $R_C(t) \sim 10^{-14.5}[(10^{19}-t)/10^{19}] = 10^{-14.5} - 10^{-33.5}t$ and $R_G(t) = 10^{-53} - 10^{-63}t$, need to be set equal to get a solution. However, it is apparent immediately that the first expression could never become less than 10^{-28} metres, while the second is clearly always below 10^{-53} metres. The two can therefore never meet, and no solution appears. In the lifetime of the atom, the gravitons – in their constrained state – are always simply too small to be perceived by the external atom.

The problem could also be tackled through a proof by contradiction. Let the initial assertion be made – which it is hoped to disprove - that the situation in which the modified Planck radius $(([mG/c^2][h/mc])^{\frac{1}{2}})$ equals the gravitational radius (mG/c^2) has no solution within the lifespan of the universe. This equivalence can only occur when $h/mc = mG/c^2$, making the Planck radius the square root of $[mG/c^2][mG/c^2]$, and leading to $G = hc/m^2$. From the equation for critical *matter* density (assuming this condition for the constrained Planck pile), the relation $G = c^2R/M$ has already been justified. Now assume that the gravitational radius grows at critical matter density, while the Compton radius grows at critical radiation density. The end of the universe will come when $hc/m^2 = c^2R/M$, when the Planck and gravitational radii are finally equal, when also the quantum and macroscopic worlds fail to be separated.

Using $R_{end} = cT_{end}$, and letting t_p denote the unchanging Planck time quantum h/mc^2 – the Compton radius crossing time - the statement $h/m^2 = Rc/M$ is translated[53] into the relation $M/m = T_{end}/t_p$. The ratio M/m is actually just N, the total number of hadrons in the universe, given that M is the total hadronic mass of the universe, and m the mass of a single hadron. The time term is the universe's age, in units of the Planck time $t_p = h/mc^2$. The Planck length is currently about twenty orders of magnitude away from any physically meaningful fundamental length.

The Planck length – as the geometric mean of the Compton radius (h/mc) of the proton and its gravitational radius (mG/c^2) - is about 10^{-33} m. Various quantum effects dominate all through the forty orders of magnitude down to the gravitational (i.e. critical) radius of the neutron, but when the condition $2mG/c^2 \approx h/mc$ is reached, the remaining Planck pile would *itself* become like a Planck particle. The particle would then become unstable, and quickly release its remaining energy.

Based on these various assumptions, the time needed to achieve this instability is $10^{42} = T_{end}/t_p = T_{end}/10^{-23}$ seconds. Hence, $T_{end} = 10^{19}$ seconds, or ~3 x 10^{11} years - *possibly* within "real"-time. This approach has therefore shown the assertion to be *false*. If the lifetime of the universe can reach, say, 10^{19} seconds – and, if it *is* legitimate to use the substitution $G = c^2R/M$, used for the critical matter density situation in the graviton pile - it becomes mathematically possible to have the Planck radius coincide with the gravitational radius. Nevertheless, this coincidence basically takes place at or just beyond the predicted age of the universe, so the age of the universe is in no way foreshortened by this effect.

---

53   $M/m^2 = Rc/h$, so $M/m = mRc/h = mc^2T_{end}/h = T_{end}/t_p$.

The relation $R_c(t) = 10^{-14.5} - 10^{-33.5}t$ – used above – is approximate. A more accurate description for the time-evolving radius of the Compton atom comes from inverting the volume equation, written as a time-varying function. Thus, $V = \dfrac{4}{3}\pi r^3$ inverts to $r = \sqrt[3]{3V/4\pi}$,

which is adapted as

$$r(t) = \sqrt[3]{3 \times (10^{-44.5} - 2.1 \times 10^{-63.822}\, t)/4\pi}.$$

and this parametric figure was fed in as the "working figure" in the calculation made earlier in the section entitled *A Creation Scenario*.

COSMOLOGICAL RED SHIFT AND THE OPEN UNIVERSE

Figure 23: More Wandjina Petroglyphs
Rock Paintings from Kimberley, Australia, estimated at 5,000 years old.

400 BC India: From the *Mahabbarata*. Blazing discs burned and destroyed an entire city and it's inhabitants, before returning to the hand of Vishnu.

329 BC: Alexander the Great records two great silver shields in the sky, spitting fire around their rims, that dived repeatedly at his army as they were attempting a river crossing. The action so panicked his elephants, horses and men that they had to abandon their river crossing until the following day.

322 BC: Seven years later, while attacking a Venetian city in the eastern Mediterranean, observers on both sides of the conflict reported another incredible event. Objects appeared in the sky. One of the objects suddenly shot a beam of light at the city wall, crumbling it to dust. This allowed Alexander's troops to easily take the city.

Cosmological red shift was the starting point in this whole discourse, as it permitted the assertion to be made that the universe is expanding. Yet, the explanation of this phenomenon has been left almost until last. The reason for this is that ultimately the proof of the current theory begins and ends in this same observation, which is nearly the only thing that can with certainty be measured and act as a guide to how the universe has evolved and changed with time.

The photon carries a record – not about its history, but about its origins – in its red shift. With modern instruments, a wavelength difference is measured between incumbent photons and rest frame reference photons. To the degree that time sped up in the local universe since the emission of the photon in *its* local universe, a relative *frequency* shift is now observed. The absolute spin rate of the photon has never changed, but the external clock *has*. The degree to which the clock has sped-up now appears as an apparent decrease in photon energy, quantified by its increased wavelength.

In general, a gravitational contribution to red shift remains virtually unmeasured, always at less than about one part in a hundred million. The only exception is for the case of light emitted from near the Quasar Era. At this time of near-critical radiation density, local clocks ran significantly more slowly and matter densities were significantly higher.

It is also possible to superimpose the peculiar and rotational velocities of the source object onto the cosmological red shift. These can either add to or subtract from the shift, depending on the trajectory of the source. This effect is more marked for photons arriving from nearby objects, as the peculiar velocity forms a relatively larger proportion of the observed spectral shift. For very distant objects, the equivalent cosmological "recession" velocity may be in the order of two hundred thousand kilometres per second, and so peculiar velocities of around 100 kilometres per second make no important difference. For cosmological recessions smaller than ten thousand kilometres per second, peculiar velocities represent an uncertainty of a few percent.

Distance measurements made at high red shift will *generally* be accurate, from the point of view that the cosmological effect completely dominates all others. Near the quasar era, however, things are more complicated. Light arriving from distant galaxies presumably left those galaxies early in local history. At that time, the universe was much smaller, and objects were much closer together than now. The Hubble Law gives a linear distance-to-red shift correlation apparently to *all* distances, so observers should expect to see a certain curvature in space (pushing objects together) as the look-back time increases. Matter densities will seem to increase hyperbolically at the horizon, giving a false impression of extra mass.

To explain this peculiar effect, imagine that the universe began with a single object. This should lie at the horizon of *every* view of space. That is, it should appear at the horizon, no matter which direction one looked. Horizon objects must therefore keep zooming around into the field of view on what seem to be hyperbolic paths, and the universe seems to operate like a giant convex lens.

Galaxy densities estimated on the basis of counts made at these large red shifts will clearly be too high. Local densities will be far more reliable, but this wrap-around effect cannot be avoided for deep observations.

A rapid drop in the number density of galaxies is predicted near $z = 0.8$, because many galaxies formed at that red shift *appear* to be at greater red shifts due to the addition of gravitational red shifting, and due also to a temporary drop in the speed of light at that epoch. The quasars and earliest galaxies will nevertheless produce more total luminosity at that red shift – despite lower galaxy counts - because of the high temperatures in that era. Beyond $z = 0.85$, the proto-galactic clouds should exist, low in luminosity, cool, and basically invisible. The red shift boundary for galaxies should lie somewhere near $z = 0.8$ (as this is when galaxies first began to form), and this will soon be within the range of telescopes. Against this, however, the convexity of the universe (the wrap-around effect) will generate the appearance of similar number densities at red shifts exceeding $z = 0.8$. This effect will need to be disentangled.

The Standard (Cosmological) Model's solution - that the universe is critically closed (i.e. has just enough mass in it – seen or not – to stop its expansion asymptotically at infinity) – is one certainly *favoured* for current mathematical models. Arguments about the so-called *acceleration parameter*, Ω, are simplest and most consistent when $\Omega = 1$, which is popularly called a "flat" universe. Basically, when $\Omega < 1$, the universe "opens up", as if it were below the critical density needed for self-gravity to halt the momentum of the expansion generated by the Big Bang. If $\Omega > 1$, the universe would have sufficient self-gravity to close itself down, and would eventually flow back in a cosmic *crunch*.

In short, Ω becomes a number that speaks about the mass density of the universe. But – in this new model - matter is responsible for *both* the expansion and the gravitational restraint seen in the universe. It is *also* responsible for maintaining the speed c at an ever-constant value. It is, therefore - *by definition of this radiation-based model* - impossible for the speed of the expansion to slow down or speed up, as long as gravitons still exist to drive it. So, the universe neither accelerates nor decelerates. Obviously $\Omega = 1$ exactly, as a basic property of this universe, and this is irrespective of mass density.

Even the Standard Model of the Big Bang implicitly suggests and prefers a non-closed universe. Every explosion known in nature is an attempt to move a situation *away* from a critical state. Cosmologists would *argue* that the Big Bang is not, in fact, an explosion, because it is not contained within any other system. Nevertheless, the thought that the universe is a system not within any other system is not yet proved: it is an assumption made in the spirit of GR that the universe is the simplest entity it can possibly be. It is the "everything" within which all else is.

However, because the *same* universe will exist until it finally sinks back into the vacuum as radiation, it seems that it is necessarily a critical *radiation* – not matter - density that finally governs its fate. It has been revealed in this discourse that the critical condition for an expanding universe - in which all atoms radiate gravitons, and in which this *radiation* is the source of the cosmic expansion - is that $R \propto \sqrt{M}$.

The total mass of the universe is 10^{81} x 10^{-27} x 10^{1} = 10^{55} kg. Critical radiation density should occur when the radius is approximately $\sqrt{10^{55}}$ = 3 x 10^{27} metres. The current radius is about 1.1 x is 10^{26} metres, which is lower than the radiation balance required for a Static Universe. The universe still needs to become thirty times older before this situation occurs, and the expansion of the universe could very well boil down to the universe searching for this situation of Static Balance.

Maximum consistency of observation with theory arises from a universe already about twenty billion years old, a Quasar and Big Bang Era starting about sixteen billion years ago, and an eventual age for the universe of up to three hundred billion years. This predicted age assumes for now that the universe does not lapse into a Steady State (radiation balanced) universe before its otherwise final demise at 10^{19} seconds. In the meantime, though, this new model has provided an entirely consistent and more "physical" interpretation of what is seen in general cosmology and empirical physics.

Is Antigravity Possible?

1947: The first major American wave of "flying disc" sightings started with formation of oval flying objects seen weaving through the Cascade Mountains of Washington State at 1700 mph, by businessman/pilot Kenneth Arnold. Ground observers reported seeing formations of discs at the same time and place. More than 1500 reports of daylight sightings were reported in newspapers. The Roswell incident occurred in the same year.

1957: Third American UFO wave highlighted by electrical interference reports around Levelland, Texas. Air Force blames the many car-stoppages on an intense electrical storm, even though the night was clear.

The term "antigravity" is used glibly in pseudo-science circles, as if the universe had a hidden force of gravitational repulsion that is just the flip side of gravitational attraction. Theories of an "anti-universe" speak of "anti-mass", "anti-light" and "anti-gravity". Not that such an idea is unreasonable. But the word "anti", when used like this, more commonly means "out-of-phase-with" than "opposite", so that it is a parallel universe that is meant. From that point of view, anti-gravity would act in the same direction as ordinary gravity, except that it would act only on anti-mass, bending only anti-light. What is needed is antigravity acting on ordinary mass.

The search for an anti-gravitational machine stems in part from the sheer technical obstacles seen in contemporary space travel. Rocket launches are always dangerous, and the need to store huge amounts of fuel for take-off always makes rockets very bulky and expensive. Rockets need to climb out of a huge gravitational well just to get into orbit. And these are small issues when compared to the problem of interstellar travel, as it takes years – even at the speed of light – to get to most stars. And what of the required fuel? Besides, rockets and satellites – fast as they are - are just too slow for practical space exploration purposes. The nearest star is more than four light-years away, and our own Milky Way galaxy is some 100000 light years across.

Having developed a new theory above to explain gravity, it could now be interesting to consider if one might, in practice, be able to manipulate space and time so as not to be constrained by gravity. Discussion will start with a search for something perhaps best called *negative mass*.

To explain this term, it is necessary to revisit the basic concepts: gravitons flowing from an atom drive the expansion of the universe, and gravitons being reabsorbed by an atom produce the force of attraction called gravity, and also assign mass to that atom. The degree to which gravitons are reabsorbed is the degree to which an "unhindered" clock – determined by the rate of emission of gravitons from a lone atom in the universe – is slowed down. When inward and outward graviton fluxes are equal, time stops completely and the atom disintegrates. Gravity can be characterized as a local "drag" in the expansion of space – it is a negative inertial effect. This negative inertial effect is equivalent to what is designated as the "mass" of the neutron. In the language of GR, this mass curves the space around it.

The easiest way to show the effect of gravity, and the situation that is necessary to overcome it, is through the diagram below. The circle represents an emitting neutron. The horizontal line represents the ideal, unhindered (zero gravity) expansion rate of the local universe, at speed c. This is the state in which the mass is measured as zero. The two sets of curves (A and B) represent the normal state (A), in which the re-absorption of gravitons produces gravity and positive mass, and the required "anti-gravitational" state (B) that results in what in effect is a *negative* mass.

Figure 24: Negative Mass

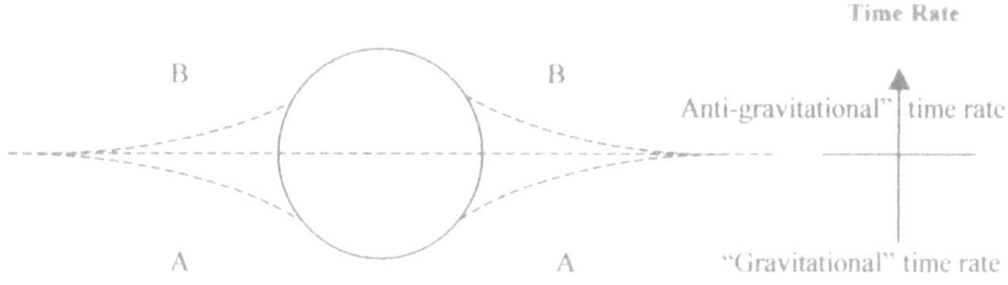

Evidently an anti-gravitational effect (i.e. local space expanding more rapidly than the general expansion) comes associated with a speeding up in the rate of time. It is the same effect on the surrounding space as if the nuclear clock had started to speed up and produce an over-abundance of gravitons. Right here there is one way to get anti-gravity – just speed up the local atomic clock! But accepting for now that this is not a viable option, it seems that a physical process must be found to create a bulge or expansion in the local bulk flow of vacuum particles, in which the gravitons are allowed to travel at speeds in excess of the average flow rate c.

The first impression upon hearing this solution is that it is impossible. Graviton emission speed c is set by the angular momentum increments given to the spirus pairs from the Planck pile at the heart of the atom. It is hard to see how this could be altered. Yet - by the same argument - there is nothing that says a graviton cannot be accelerated or decelerated once in free space. Indeed, when photons pass through materials with refractive indices $n > 1$, the speed of transmission appears to fall to below c, and then climb back to c when the photon re-enters free space. The emergent photon simply assumes the fastest speed it can, which is the speed of the vacuum flow.

ELECTROSTATIC FIELDS AND GRAVITY.

The earlier model for the six quark loops within the neutron, before the final emergence of the graviton, has the quarks changing their charge at each cycle, and reaching zero charge only at the seventh cycle – the cycle seeing the release of the graviton at speed c. But what if the charges associated with each quark are also an indication of speed? It could mean that those quarks with, for instance, a negative charge could be travelling in excess of c (hence n < 1), while those with a positive charge travel slightly slower than c (hence n > 1), and the emergent charge-less quark emerges at exactly c (thus, n = 1)! Such a model to explain why a charge causes a "stress" in space is certainly believable. It should also be testable.

The quarks would then cycle between speeds above and below c in a regular fashion. Given that a full oscillation takes about one Planck time, a ripple frequency of about 10^{23} Hertz should appear in the atom. This ripple speed would define an associated (de Broglie) wavelength of $\lambda = c/f = 3 \times 10^{8}/10^{23} = 3 \times 10^{-15}$ metres, defining the Compton radius. From $E = hc/\lambda$, it follows that each neutron represents some 2×10^{-10}J of energy, or an associated mass of $E/c^2 = 2 \times 10^{-27}$ kg – which is, indeed, about the mass of the neutron.

In standard theories, the speed of light is regarded as being governed by quantities like the permittivity and permeability of free space (ε_o = 8.85

x 10^{-12} and μ_o = 4π x 10^{-7}), which are electric and magnetic constants describing the level of interaction of photons with the neutral vacuum. Specifically, c^2 = $1/\varepsilon_o\mu_o$ in free space. These constants specify the speed of light in terms of conformity to the laws of electromagnetism and the properties of dielectrics. The electric force is much stronger than the gravitational force, and is also inverse-square.

In terms of energy, E = mc^2 leads to the result m = $\varepsilon_o\mu_o$E. So, if the values of these free space constants decrease, so does the mass associated with a specific energy. When one or other constant is zero, mass becomes zero. These numbers are very small in any case, as the vacuum interaction with photons is extremely tenuous. The case of ε_o = 0 is that in which atoms absorb no gravitons at all and the atomic clock runs at its unrestricted rate. When any matter is present, the clock slows slightly, and ε_o gains a small positive value. Space curves back toward that mass concentration.

But in the presence of a negative electric charge, the vacuum swells out much more than gravity bends it in, and the balance set up between the two effects sees ε_o attain what must indeed be a small *negative* value. A sign change has therefore ensued, and this would effectively produce a "negative" mass – where an object's mass is effectively projected back onto its surroundings. This causes not just a decrease in the mass associated with some system energy, but an actual "anti-mass" effect in the particles producing this effect. This is balanced, overall, by positive charges increasing the value of ε_o in an opposite way, and slowing the clocks down more than is normally attributable to the gravitational effect. Positive charge adds an equal amount of positive mass to an object. On average, ε_o and μ_o have the values associated with them, as long as they are far from any electric charges. But this may not apply accurately near static charges, where the vacuum is pre-stressed.

Here, then, is one possible way to distort space and cause a "negative mass" effect. Create a growing, massive negatively charged region above a vehicle, and a neutral or oppositely charged region below it! These regions need to be arranged so that the pinching of space is *behind* the motion, and the swelling *ahead* of it. The large negative charge sphere around the small positive atomic nucleus suggests that probably it is the negative charge that swells the vacuum and the positive that pinches it. The two charged regions do not even have to be isolated, as long as they continue to be sustained. But they cannot be allowed to dissipate or merge.

By having a swelling region in front of the vehicle and a pinching one behind, there is a kind of "engine" that pushes the craft forward through the vacuum. But there is no change in the overall mass in this configuration, as the appearance of negative mass at the leading edge is matched by the increase of positive mass at the trailing edge. The sum stays fixed. By having a neutral region behind the craft, there is at least no positive mass gain at the trailing edge but there is only half of the forward drive on the vehicle. But if this comes with the addition of negative mass on the leading side of the craft, the loss of some thrust does not matter - the craft will still go forward.

As the charged clouds become smaller or weaker, the local distortion in time will also weaken, and the vehicle will begin to lose its "negative mass", becoming more and more "massive". The reappearance of mass will feel like a strong deceleration. But - while the imbalance exists - the central vehicle will feel a kind of "pressure", as the vacuum particles behind the craft push it forward into the void in front. The vacuum urgency to even out the flow will drive the vehicle forward.

Figure 25: Relative Spacing of Gravitons in Charged Vacuum

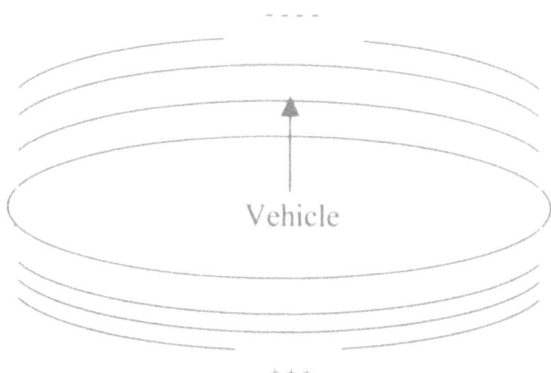

Another way to look at it is that the clock in front of the vehicle runs a little faster than that on the vehicle itself, which in turn runs a little faster than that behind the vehicle – if a positive charge cloud is also located there. A differential in time is equivalent to acceleration, which itself is physically indistinguishable from a "force of gravity", acting in the direction of the differential. This differential can be pointed in any direction. In the case of producing just a negative cloud in front of the craft, but a neutral one behind it, it becomes necessary to separate the elementary charges first, and then to expel one type of charged particle while accumulating the other. The craft therefore needs to expel a positive-ion exhaust trail.

What would be useful in building such a vehicle would be to find a substance into which energy could simply be supplied at a steady rate, while on its front surface electric charge continues to accumulate. Some kind of super-diode might be in mind. If the same process simultaneously expelled charge from the rear of the vehicle (while maintaining the charges in separation), it would in principle be possible to build up enough localised charge to produce a "significant" stress in the vacuum ahead of the craft. The question then arises as to just how much of this stress is needed, and how much can be created, per kilogram of vehicle mass, to affect a useful anti-gravitational force.

A "charged region", in the description above, does not necessarily have to mean charged particles, although this is the natural choice. It means any process that has the same effect as an extremely large concentration of static charge ahead of the vehicle. Plasma, and its localization, seems to present one natural choice. Plasma is formed artificially when suitably doped materials are bombarded with laser light, inducing them to eject their electrons *en masse*. A mechanism could focus on immediately isolating these electrons and expelling the leftover positive ion plasma at high velocity. Confinement rings would circulate separated electrons and positive ions in opposite directions, keeping vehicle rotation as close as possible to zero. Expelled plasma could perhaps be focused down to a pencil-like beam before being ejected.

One further possible way to screen the craft from external gravitons is to allow the accumulated electrons to circulate rapidly in front of the craft to create a vortex, partially deflecting vacuum particles into a spiral around the craft. The confinement ring for the electrons would need to be placed ahead of the craft. A vortex would amplify the anti-mass effect and allow a faster-than-c propagation through the vacuum. The craft would receive impetus by creating a huge rotating - but still generally static - voltage on the vacuum before it.

Static voltage (measuring potential energy in an electric field) obeys the rule that $V = Ed$, where E is measured in Newtons of force per Coulomb of charge (N/C). A craft of 10^3 kg would have a weight of about 10^4N. Levitating such a craft requires a balance between weight and electrostatic force, or $mg = qE$. A moderately strong electric field of 10^4 N/C, acting against just a Coulomb of charge (6 x 10^{18} charges – which weighs only 10^{-12} kilograms) will produce this balance. It is potentially a strong force.

Consider a pulsed laser that creates 10^{-3} kg of plasma per second by vaporising fuel pellets. The plasma laser would be run by a small nuclear reactor, or the like, of by re-tapping energy from the plasma before ejection. If positively charged plasma is ejected as fine vapour, with exhaust velocity 10^5 m/s, there will be a momentum transfer of 1000 kgm/s, accelerating the 10^3 kg craft at 1 m/s^2. After a year of converting plasma at this rate, the craft could reach about ~0.1c from momentum transfer alone. This is a speed that could make interstellar travel possible, but the problem is that the plasma engine will have ejected some 3 x 10^4 kg of mass in this time – more than the mass of the craft!

Here, the electrically charged leading surface, and a vortex effect, could begin to make a real difference. Indeed, once the leading surface acquires enough "negative mass" to balance the conventional mass of the whole craft, the mass will effectively be zero, and the craft will have reached the speed of light. Indeed, with the vortex present, it will be possible to go even faster than this, but relativistic effects will appear near c. Physically, this is a situation where gravitons can no longer find their way through the craft's dense electrical shield so as to contact the atoms from which the craft is constituted. However, before this situation can arrive, the vacuum will have begun to build a shock wave against the charged leading surface of the craft, fighting to compress the swelling caused by the accumulated charge. Charge density will climb steeply, and a physical cap may be set on the charge that can be aggregated. The growth of negative mass will tail off as the craft becomes relativistic.

The physical description above suggests an equation for mass similar to the relativistic one, but with a correction for negative mass in the rest mass term. The fact that collections of electrons generate some "negative mass" may mean the usual starting value for electron mass needs first to be corrected upwards. The usual value is 9.11×10^{-31} kg. For this calculation, the mass of the electron will be defined as $m_e = \varepsilon_o q_e$ = $8.85 \times 10^{-12} \times 1.67 \times 10^{-19} = 14.8 \times 10^{-31}$ kg, a number suggesting that an electron may appear some 38% less massive than it really is because its charge acts to distort local space. By the same token, the proton should seem about 38% more massive than it really is, so a "charge-less" equivalent mass of 1.21×10^{-27} kg is suggested for the 1.67×10^{-27} kg found experimentally. The true mass ratio between proton and electron may therefore be 818, rather than the conventionally accepted value of 1836.

The negative mass contributed by the electrons increases, as the proton plasma is ejected. The electrons cause a 38% effective loss in all of the mass they hide from the vacuum, and not just in themselves. The proportion denoted by the mass consumed through lasing and expulsion can be denoted as p. Thus the vehicle mass at any particular time is denoted by $M_o(1 - p)$, and electron mass is $M_o p/818$. The "negative mass" generated by this cloud of electrons at rest is $0.38[M_o(1 - p) + M_o p/818]$. The negative mass situation should also see a faster propagation speed at any time, in accordance with p. The vortex effect can be factored in later.

The relativistic mass - at such time as a proportion p of the craft has been consumed for fuel – is

$$M = (M_o(1 - p) - 0.38[M_o(1 - p) + M_o p/818])/\sqrt{(1 - (v/(1 - p))^2/c^2)}$$
$$= (0.62 M_o(1 - p) - 0.38 M_o p/818)/\sqrt{(1 - (v/(1 - p))^2/c^2)}.$$

The acceleration function for this craft is simply a = F/M. The ratio of accelerations for craft with and without the "negative mass" effect is

$a_{nm}/a = M/M_{nm}$

$= (M_o(1-p)/\sqrt{(1-v^2/c^2)} / (0.62M_o(1-p) - 0.38M_o p/818)/\sqrt{(1-(v/(1-p))^2/c^2)}$

$$= \frac{M_o(1-p)\sqrt{(1-(v/(1-p))^2/c^2)}}{\sqrt{(0.62M_o(1-p) - 0.38M_o p/818)(1-v^2/c^2)}}.$$

If $M_o = 10^4$ kg, and the consumed mass fraction reaches $p = 0.1$, the ratio is numerically

$$9000\sqrt{(1 - 1.23v^2/c^2)} / 5576\sqrt{(1-v^2/c^2)} \approx 2.$$

The acceleration is thus approximately doubled by the negative mass effect. If p is moved up to 0.5, the acceleration ratio stays about the same.

One can only surmise as to what the effect of the vortex might be, but perhaps it could again double these gains only by increasing spin alignment of the vacuum particles much further ahead of the craft than is usual. To create such a swirling electron cloud, magnetic fields are required. These confine and circulate the electron beams. If pulsed, they may be able to push electrons up from the surface of the craft. The craft may have to counter-rotate somewhere to conserve angular momentum from bulk-flows of circulating plasma, but it should not be a major effect. With a vortex present, the craft may also stay quieter within an atmosphere, as the compression wave - in the direction of motion – is what produces the sonic boom that usually develops as a craft attains the speed of sound.

The time it takes to get to a consumption level of $p = 0.5$, and the speed the craft attains by then, will depend upon the efficiency of the plasma engine and the integration and subtleties of the craft's propulsion systems. On first principles, however, if the vehicle consumes 0.17 gram per second, for a year, a total of 5 x 10^3 kg of plasma pellets will be consumed. This represents about 50% of the 10^4 kg craft, or $p = 0.5$. If an acceleration of 2 m/s² were achieved throughout, a minimum speed of 6 x 10^7 m/s, or $0.2c$, would be reached in this time. After two years, a speed of $0.4c$ is possible, and a distance of 0.3 light-years covered.

With the vortex factored in, the speed might by this time get up to $0.8c$, and a final speed near c might be attained. At speeds below $0.9c$, the relativistic effect should not be significant, and - with the vortex operative - a speed nearer c may be possible before the craft becomes relativistic. To cover the journey to the nearest star and back might take as little as ten years.

Creating plasmas is technically possible. Electrons associated with metals and their alloys typically resonate in the microwave part of the EM spectrum. Plasmas generated by millimetre band lasers ($10^{-1} > \lambda > 10^{-3}$ m), in particular, have two appealing properties: if applied in surface-wave sustained mode, it is possible to generate distended plasmas of high density. In addition, these plasmas exhibit a high degree of spatial localization and can be accumulated without rapid charge loss. They can therefore be rotated, to create electrical vortices. Atoms resonating in the millimetre-wavelength band release electromagnetic waves in the frequency range 3 – 300 GHz, and this should produce a detectable signal. Millimetre waves are found at the long end of the microwave spectrum.

One problem with the scenario in which separated charge clouds are found at *both* ends of the craft is that the craft begins to look like a giant capacitor. The use of a capacitive vehicle would always require a way to overcome the dielectric reaction that occurs across the middle of the craft. Living quarters would need to be made utterly non-conducting, non-magnetic and non-dielectric. But materials exist today whose properties change as different electric fields are passed across or through them. Ferromagnetic substances can be rendered virtually impermeable to magnetism with the right doping by other atoms and a careful production process. Graduated doping could allow material properties to change even in going from surface to interior, allowing both surface dielectric function and the necessary interior insulation.

REFRACTION AND GRAVITY

Earlier, the idea of accelerating the atomic clock was mentioned, but nothing more was said. The reason is that nothing, according to the current theory, should exist that could either speed this clock up or slow it down. This clock manipulation was needed to get gravitons to leave atoms at speeds other than c. Yet there is already a physical phenomenon that appears to see the speed of light (and, therefore, of gravitons) altered, *after* those gravitons have been released into the free vacuum. The refractive index, n ($= \lambda_{vacuum}/\lambda_{medium} = c_{vacuum}/c_{medium}$), conserves photon energy both in the vacuum and in any other transparent material medium, so that $E_{vacuum} = E_{medium}$. This implies that $hc_{vacuum}/\lambda_{vacuum} = hc_{medium}/\lambda_{medium}$, or $c_{vacuum}/\lambda_{vacuum} = c_{medium}/\lambda_{medium}$. That is, $c_{medium} = c_{vacuum}/n$. Frequency is observed to be the only invariant in all media, indicating that it is an *intrinsic* property of photons. The relation $c = f\lambda$ confirms that if λ changes, c must do so in direct proportion, as f is fixed. And this is observed.

Huygens explained, with his diffracting photon wave fronts, why refraction occurs. Part of a wave meets a slower medium, while part is still in the faster medium, and the wave front rotates because of the differential speeds. The wave fronts crowd together in the slower medium, and the wavelength is reduced. He assumed, however, that the wave was propagated linearly through the second medium, and the inferred ratio between the incident and refracted speeds of light derives from this. Huygens did not, however, venture to explain what determined this new propagation speed.

Currently no textbook theory teaches *why* – physically - the speed of light slows in the presence of matter, to give rise to the quantity *n*. Textbooks simply supply tables of values for a range of materials. But it is *not*, at least, considered to be a relativistic effect. Checking for the constancy of n when the refractive medium is subjected to relativistic speeds could prove to be an informative experiment.

However, there does seem to be some link to atomic spacing or configuration. Strongly bonded substances – like diamond – show the largest values of *n*. Amorphous substances like water and glass have smaller refractive indices, and if glass is doped with atoms to make it harder (and more crystalline and atomically regular), it's index climbs. The refractive index of air is also temperature and pressure dependent – that is, varies with density. The twilight sun is seen when it is already over the horizon because the atmosphere bends the light over the horizon – an effect due to the different densities near the ground and further up. As air density increases, refractive index climbs, and this is true for other substances as well. Beyond an index of about 1.6, photons will simply not be transmitted at all, and are reflected completely, or absorbed.

Not all materials transmit light, of course. Most substances reflect or absorb, and it is from the differences between reflection and refraction that a better understanding of refraction arises. Photons are not solid, and do not "bounce" from atoms as if atoms were solid objects. In reflection, surface electrons simply absorb the incumbent photons by assuming excitation states. They also receive a transfer of angular (but not linear) momentum from the photons. Photons emerge at the equal reflection angle simply because when the angular momentum is restored in the form of a photon (during electron de-excitation) it seeks to balance the angular momentum being received from a second incident photon. With a beam of light, the photons arrive in a regular manner, and set up this balance.

The angle of incidence therefore mirrors the angle of reflection. If photons arrived sporadically, the angle of reflection would be random.

It is the electrons found orbiting the atoms in the top few layers of a "solid" substance that do all of this. Few photons get past them, if the solid is reflective. Random thermal motions of atoms or crystal lattice imperfections may allow partial transfer and partial reflection, but in the absence of these causes all photons would either be reflected or transferred, depending on the material. Mixed reflection and refraction in amorphous substances like glass stems from the absence of a regular crystal lattice.

Transmission takes place when photons – managing not to be absorbed by surface layer electrons - take a kind of oscillating (whether spiralling or semi-random) path through the lattice. The photons find their way in the flow of vacuum particles through the spaces between the atoms, but this flow alternately swells and pinches, causing accelerations and decelerations of the photons. The photons may also pass to and from atoms as they pass them by. The collective effect of these variations is to maintain an average speed of c, but the longer total path is seen externally as a slow-down of the light-speed within the medium. The path length through the atomic array is now n times longer than the direct physical dimension of the substance, but the photon still travels and emerges at the speed of light.

It seems likely that the linear speed of photons in a refractive medium is not reduced at all. This seems all the more clear for another reason: had the speed of light changed, a stress should have emerged in the vacuum encased by that substance, and an electric or magnetic field should have appeared, to reflect this. But this is not observed.

The gravitons that enter the atomic substance share a slightly different fate. They do not enter in coherent beams, and follow far more random paths through the structure. Many are absorbed into atoms, and removed from the vacuum. This interaction accounts for the "mass" of the refractive material.

The conclusion appears to be that the electron shells of atoms are not involved in slowing the atomic clock, so that an incident light ray is not slowed in a refractive medium because of the slowing of atomic clocks. Such a slowing of clocks can only occur when more gravitons are absorbed because of the congestion of matter. The effect is unchanged by the presence of photons. It seems that the creation of substances in which the refractive index can be varied will not pay a dividend as far as helping the craft go any faster.

Indeed, there are relatively few basic processes in the universe that can be manipulated to build such a vehicle. And it seems that refraction is not one of them. Only the electric force can distort the vacuum in a way that looks like a gravitational effect. The "electro-magnetic" propulsion option for the vehicle still seems to hold the most promise.

A CRAFT THAT DEFIES GRAVITY?

Figure 26: Dreams of Other Worlds.
A drawing from a philosophical novel named *Micromégas* (1752), by the French writer Voltaire (1684-1778).

One advantage of a plasma-capacitor-style vehicle would be the possibility of keeping it so much smaller than a conventional rocket. But, then again, there is also no urgent need to make it a small vehicle, unless it needs to be parked in confined spaces or has problems with structural integrity. The craft needs energy to run a pulsed laser, energy to create magnetic fields to confine plasma, and energy to spin confined electric fields into vortices. Some plasma energy can be re-tapped to run simpler functions, such as guidance, heating, cooling, communications, etc. A small atomic plant can run all else, and act as a backup. There is no need for vast reserves of compressed liquid or gas fuels, which – apart from their excessive weight - occupy an ungainly volume, and determine the conventional "rocket" shape. A rocket is designed both to contain fuel and to be aerodynamic, and hopefully an ideal interstellar craft can get away from this ungainly shape and size.

The craft being imagined would create its own sense of weight, as – after all – weight is just the effect of the ground withstanding downward acceleration. As the craft accelerates, any object on its floor will feel a "weight", and this may limit its possible acceleration rate. When "coasting", all becomes weightless. In this latter situation it may be helpful to give parts of the craft the ability to rotate and create artificial gravity. The rotation of these sections could be driven by magnetism. Magnetic cushions under the floor could also soften the g-forces associated with acceleration.

For landing, the craft needs to have the ability to hover down smoothly and "park" itself. The "negative mass" needs to decrease to zero - and the vehicle even needs to attain a slight "positive" mass - to allow the craft to alight. Legs are optional, but if a power plant is attached beneath the craft this will need to be protected. Perhaps the craft could retract this power unit up within itself at landing, and lower a cylindrical ring instead, to act as its base. Ideally, in the final stages of touch-down, it may be able to utilise a second set of plasma lasers to superheat and ionise air immediately below the craft, creating some "cushion" of hot expanding air to help it down.

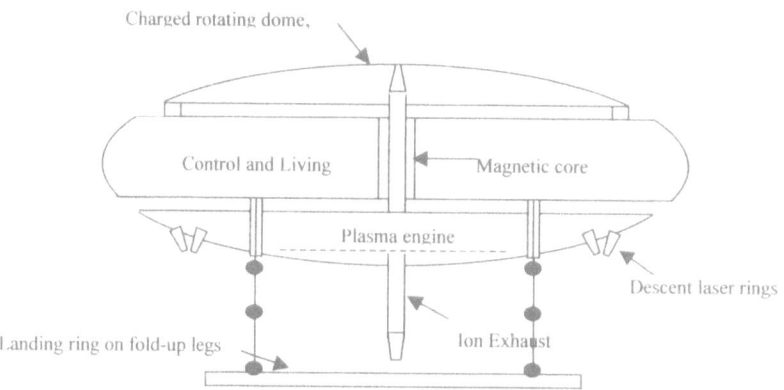

Charged rotating dome,

Control and Living

Magnetic core

Plasma engine

Descent laser rings

Landing ring on fold-up legs

Ion Exhaust

The popular belief that such a craft should be disk-shaped is well founded. A disk is certainly aerodynamic, yet this would not be the main reason for the shape. Balance and control are more important. The vehicle needs no front or back, but it does need a sense of forward and backward. It will probably need the ability to tilt the driving mechanism so that the vehicle could hover or move sideways. An almost certainly circular confinement ring for plasma electrons further suggests a circular craft. Retractable landing gear is best if circular. And a circular shape is convenient for providing the craft with a protective circum-vehicular magnetic shield.

There is no need for highly reinforced structures to withstand vibration, impact or relativistic particles, except at the slowest speeds. The same magnetic fields as are used to confine plasma while the craft is accelerated to its high interstellar speeds can afterwards (when the craft coasts) be used as a shield against potentially dangerous particles (in the same way as the Earth's own magnetic field protects its inhabitants from the dangers of solar radiation). High-speed travel might involve relativistic changes, but these are not structural stresses. Stresses experienced at high speed in the vacuum are the same as those at lower speeds. It is not the same as travelling in air, as resistance is not felt from the external medium, except near the Earth, where velocities must of course become non-relativistic. Radar would be needed to protect against possible collisions with free objects in space.

The key requirement is to find an electrically and magnetically impermeable, shatter- and vibration-proof, heat resistant, "light" material that maintains reliable structural integrity. The required material may, indeed, be non-metallic. A super-plastic, or perhaps a plastic sealed over a steel-alloy net, sound like good contemporary possibilities. Strong magnetic effects created by such a craft might distort incoming light or interfere with communications. The viewing problem requires compensating magnetic fields within the vehicle's viewing windows; communications problems have to do with finding transmission frequencies that circumvent interference.

Sufficiently powerful magnetic fields can act like a distorting lens, bringing background light around in front of a craft, partially obscuring it. An outside observer might find the craft hard to see, as fewer photons reach the surface of the craft so as to be reflected from it, and when reflected these photons are also bent away from the viewer. At relativistic speeds, such a craft would be virtually invisible from the side. The craft would, however, be visible whenever it was viewed in a "top- down" view. In this configuration there should also be polarisation rings seen around the vehicle.

If charged particles interact with these powerful fields, they find themselves spiralling around the field lines and bouncing between the magnetic poles, possibly emitting light as they do. This could make the poles of the craft visible. Gas dampers at the poles could induce de-excitement of these charged particles, rather in the fashion of charged particles interacting with the Earth's atmospheric gases to produce the aurora.

Figure 28: Magnetic Shielding of High-Speed Charged Craft

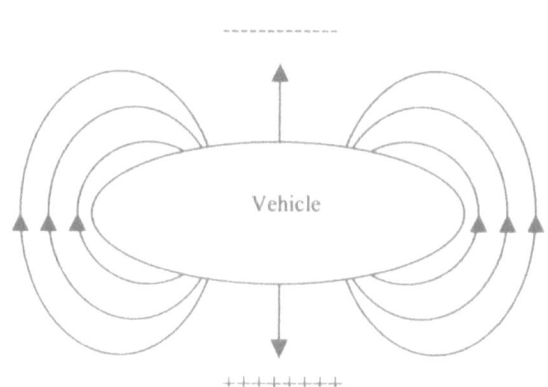

Another thing about a charged-sphere, plasma-based craft is that several such craft can be stored together in a rack-like fashion within a larger craft, such as a cylindrical "mother" ship or station. Within this rack, these craft could perhaps be repaired, "discharged" or even have their power supplies replenished or replaced. At the very least, these smaller craft could be transported in numbers into space, before being sent out on individual missions. The "mother craft" would contain all the materials needed to maintain and protect the smaller craft. It could remain "parked" in space, while the disk-like craft went out on specific missions. Both mother and daughter craft would generally avoid the need to "land" anywhere.

For the rest, the construction of such a vehicle is over to the experts. Rotating sets of magnets, magnetic pinching and confinement rings, pulsed laser plasma production, atomic power plants, magnetic shielding and computerised systems and control. It is all now a technical task. But the broad principles of operation and design are here suggested, developed from a purely logistical point of view. There are reasons to believe it will work, if sufficiently powerful concentrations of charge can be developed and contained, and if vortices can be produced and sustained. Creating comfort, a healthy living space, safety against lightning strikes and meteors, and the like, form a second set of technical requirements.

This is perhaps not technology that will provide an alternative for the family car, but it certainly has possibilities for rapid interstellar travel. Most of all, it is something that has been imagined, and that is consistent with the workings of the universe. It is the product of a few hundred years of rapid advances in knowledge and technology. Making a craft like this is no longer science fiction, but an event waiting to happen, and is undoubtedly about to become a part of human intellectual property.

APPENDIX

For the Religious Philosopher

For those who are religious, there is a close correspondence between the numbers seen in the section describing the physical form of the atom (*What is the Atom?*) and those encountered in biblical apocalyptic. Coded within the Bible, six is the number of Man and the World of his physical experience. It is also the number of the creation itself. One (1) is the number denoting God as source (and common factor) of all existence. Eight appears as the number of resurrection, and denotes regeneration of the cycle of life, and seven is the number representing the driving force of the Spirit of God in human affairs, and the appearance of divine light in the world.

In the model of the atom presented in this book, the first spirus loop to spontaneously arise from the Planck pile sees the sudden emergence of a quark pair, and with it the appearance of charge and energy, and this corresponds to the role of God as Creator of (to quote the Bible) *what is seen* from *what is not seen.* The two spirii (as a quark) seem to arise from nowhere, appearing suddenly in the external atom for six ever-more-energetic cycles. Observers equate the collection of these six quarks with the atom, which is the substance of the physical universe. Thus the atom carries the "six" of creation within its own constitution and mechanism.

The seventh loop of the quarks is also, in reality, part of the physical existence of the atom, but is not seen when the atom is destroyed, as it emerges from the atom as a charge-less and mass-less graviton. This seventh pair ends up driving the expansion of the universe, creating gravity, and accounting for most of the "invisible forces" (the "laws") of the external universe. Clearly it corresponds extremely well with the Bible's "invisible spiritual forces" that drive both physical and spiritual universes, symbolised biblically in the number seven. This seventh quark can also be energised and become a photon – the physical form of light in the universe. The Spirit of God – always characterised in sevens - is also seen as the "carrier of spiritual light" into the world. The eighth cycle of the atomic machine sees the arising of a new quark from the Planck pile. The fact that an eighth cycle corresponds to a repeat of the first speaks of the cycle of regeneration.

It will seem to the religious that there is no accident that Nature should bear these very numbers. It simply confirms that the same God who reveals Himself in the Bible, and who underlines his actions symbolically in the Bible numbers, is the same God who made and sustains the universe that Man knows.

The scientist will no doubt contend that the existence of anything can only be proved by its physical presence, the ability for it to be measured and quantified, its conformity to logic and its ability to be rationally explained. This book has accordingly attempted a "rational" explanation of the universe that Man observes. It talks about things that are present, able to be measured and quantified, logical in structure and relationship, and rationally explicable. In that sense, this discourse has been "scientific". It boils the universe down to a few intrinsic properties, and how these determine the combinations and properties of the vacuum particles that are assumed to fill space.

The universe runs on miniscule, interacting, spinning entities, here called "spirii", that fill an otherwise "empty" space. There is a physically "closed universe" of these entities, but this universe exists within another "space" – not necessarily "dimensional" - which makes up the remainder of the volume seen in space. This volume is *not* part of the physical universe, and is therefore *non*-physical. This is hard to understand, but is apparently the reality. The physical universe has no "nothing" in it, but it does exist within a greater space. That greater space is therefore not "physically" a part of the universe. It is simply the non-physical "medium" within which a physical object – called the universe – can be expressed.

If the space in which the physical universe exists is not itself "physical", it follows that it must be something that – for want of a better name – constitutes the "spiritual" universe. The spiritual universe does not need to be a parallel universe – it is simply the "rest" of the universe. This "rest" is of a substance and nature that is not expressed in the "laws of physics" that man has understood and quantified. It is infinitely near to the human world, yet utterly separated from it.

The apostle Paul spoke of God as the one in whom the human being *lives and moves and has his being*. This may just be using language from the physical reality as an analogy for the human spiritual state, but it is certainly interesting that Paul makes this particular analogy between the being of God and the space within which physical reality is expressed. Why would Paul make this analogy, unless he believed that the space in which the universe evolves *is* the same space as is occupied by the Spirit of God? That suggests that the "space" within which physical existence is expressed is just the spiritual substance of God himself!

A person may argue against the existence of God on the purely physical basis of the inability for God to exist in all times at once. Scientists seem to believe that Christians believe that God runs on an "absolute clock" – an idea that is disqualified within the tenets of quantum theory. Time runs at an infinite number of local speeds, depending on where one is stationed in the universe. If God is aware of all beings and events in the universe, and is equally amenable to all beings and objects in all their states, then He exists and operates in an infinite number of time frames at once. Physically, this is not possible. No "mind" – attached to a particular time frame - could experience the realities of both a relativistic traveller and observer at the same time. More the difficulty for an infinite number of reference frames.

Yet – if the language of the Bible is accepted as inspired – God knows when a sparrow dies and knows the number of hairs on a person's head at any one time. He has an infinitely ordered intelligence that can only stem from seeing the universe either from one clock or from no clock at all. The latter option seems more in accordance with the infinite life attributed to God. Only things that follow clocks experience the human condition called "aging" - an illusion shared by all physical things. But time does not flow in the vacuum – it does not even *exist* there! It is only a reality in a decaying physical universe.

Without the existence of time, even the experience of *knowledge* must be different. Knowledge in the physical world is always expressed in terms of *process* and *fact*. By combining process and fact, the human being imagines, considers, invents and analyzes. This book is a product of processes and facts, imagination and analysis, as is all of science. The empirical method becomes nothing more than an attempt to get the facts "right', so that the combination of process and fact will become more fruitful. The human being then attempts to manipulate his physical world, to conform it to what he prefers it to be. This is generally linked to his own physical needs and susceptibilities. His personal world is driven by the internal desire to relieve the stresses in his life.

The spiritual universe does not exist in time and space, nor know death and weakness. That universe does not evolve in any selective or "stress-relieving" manner. It already exists in perfect unshakeable harmony. As in the biological universe, its beings are perfectly designed for the function they perform there. In the spiritual universe the process of natural selection is not required. All beings exist as *part* of the Spirit of God, rather than within-but-independent of it. Knowledge is not therefore possessed, acquired or stored, but is actually part of the being itself. All things are entirely instinctual and right, and the being is – in a sense - an embodiment of all relevant knowledge.

It is in this sense that God can know all things, at all times, in all places. The existence of time in the physical universe, and the fact that human beings have to *learn* all things experientially and intellectually, is what fosters disbelief in a God who could see every event in the universe on its own clock and not feel a fracturing of His consciousness. The consciousness of God is not the same as the weak consciousness of the created being.

The spiritual notion of "distance" is also altered with the concept of time. Distance, as a word, implies time, effort and speed. But, in the realm of spirit, "distance" is replaced by "place". God is not in one place as opposed to another. There are no "distances" from God. There are only places within God. In the Bible at least two occasions are recorded where distances are covered instantaneously – one when Jesus walks on the water across a stormy sea, and as he boards the ship it immediately jumps across the remaining distance it had to cover, and the crew can almost immediately disembark. The second occasion (Acts 8) is where the disciple Philip is suddenly transported at once from a baptism on the Gaza road south of Jerusalem to a town called Azotus (near modern Tel Aviv) – some fifty miles away. That is, his *physical* body is moved by spiritual power. The *physical* limits of this world can be circumvented in the timeless realm of the Spirit of God.

The true scientist cannot draw a conclusion from the physical universe about the existence of God. The physical universe operates on physical principles, creates physical effects, and dictates the physical conditions of human life. It betrays nothing of the existence of the spiritual universe. However, the physical universe has had a point of origin. The spiritual universe has no such point of origin. Clearly, the causality is "one way". The physical universe – as a "created" and decaying entity – *must* have emanated from the spiritual universe, and not vice versa. For this reason one would expect to see some parallels between the realities of the physical universe and that of the spiritual universe.

The revelation of these parallels must originate from the spiritual world and flow to the physical world. The spiritual world can access the physical at will, but the same is not true in reverse. It is not unreasonable to expect that a superior intelligence would communicate the spiritual truths through analogies with physical truths. For this reason it makes sense that the numbers found associated in this book's theory about time, gravity, light and the atom should mirror those seen in connection with parallel spiritual realities in the Bible.

www.ingramcontent.com/pod-product-compliance
Lightning Source LLC
Chambersburg PA
CBHW032002170526

45157CB00002B/505